Integrating Food Science and Engineering Knowledge Into the Food Chain

Series Editor
Kristberg Kristbergsson

More information about this series at http://www.springer.com/series/7288

ISEKI-Food Series

Series editor: Kristberg Kristbergsson, *University of Iceland, Reykjavík, Iceland*

FOOD SAFETY: A Practical and Case Study Approach
Edited by Anna McElhatton and Richard J. Marshall

ODORS IN THE FOOD INDUSTRY
Edited by Xavier Nicolay

UTILIZATION OF BY-PRODUCTS AND TREATMENT OF WASTE IN THE FOOD
INDUSTRY
Edited by Vasso Oreopoulou and Winfried Russ

PREDICTIVE MODELING AND RISK ASSESSMENT
Edited by Rui Costa and Kristberg Kristbergsson

EXPERIMENTS IN UNIT OPERATIONS AND PROCESSING OF FOODS
Edited by Maria Margarida Vieira Cortez and Peter Ho

CASE STUDIES IN FOOD SAFETY AND ENVIRONMENTAL HEALTH
Edited by Maria Margarida Vieira Cortez and Peter Ho

NOVEL TECHNOLOGIES IN FOOD SCIENCE: Their Impact on Products, Consumer
Trends, and the Environment
Edited by Anna McElhatton and Paulo José do Amaral Sobral

TRADITIONAL FOODS: General and Consumer Aspects
Edited by Kristberg Kristbergsson and Jorge Oliveira

MODERNIZATION OF TRADITIONAL FOOD PROCESSES AND PRODUCTS
Edited by Anna McElhatton and Mustapha Missbah El Idrissi

FUNCTIONAL PROPERTIES OF TRADITIONAL FOODS
Edited by Kristberg Kristbergsson and Semih Ötles

FOOD PROCESSING
Edited by Kristberg Kristbergsson and Semih Ötles

APPLIED STATISTICS FOR FOOD AND BIOTECHNOLOGY
Edited by Gerhard Schleining, Peter Ho and Severio Mannino

PHYSICAL CHEMISTRY FOR FOOD SCIENTISTS
Edited by Stephan Drusch and Kirsi Jouppila

PROCESS ENERGY IN FOOD PRODUCTION
Edited by Winfried Russ, Barbara Sturm and Kristberg Kristbergsson

CONSUMER DRIVEN DEVELOPMENT OF FOOD FOR HEALTH AND WELL
BEING
Edited by Kristberg Kristbergsson, Paola Pittia, Margarida Vieira and Howard R.
Moskowitz

BOOK ON ETHICS IN FOOD PRODUCTION AND SCIENCE
Edited by Rui Costa and Paola Pittia

Rui Costa • Paola Pittia

Editors

Food Ethics Education

 Springer

Editors
Rui Costa
CERNAS
College of Agriculture of the Polytechnic
Institute of Coimbra
Bencanta, Coimbra, Portugal

Paola Pittia
Faculty of Bioscience and Technology for
Food, Agriculture and Environment
University of Teramo
Teramo, Italy

Integrating Food Science and Engineering Knowledge Into the Food Chain
ISBN 978-3-319-87858-4   ISBN 978-3-319-64738-8   (eBook)
https://doi.org/10.1007/978-3-319-64738-8

Printed on acid-free paper

This Springer imprint is published by Springer Nature
The registered company is Springer International Publishing AG
The registered company address is: Gewerbestrasse 11, 6330 Cham, Switzerland

Preface

Education standards for food studies have been only relatively recently been developed thanks to the activities carried out within networking projects funded within the Erasmus framework, and all include a focus on ethics.

Since the start of the FOODNET project (1998–2001) to the ISEKI_Food-4 (2011–2014) Erasmus Thematic Networks, despite the recognized importance, scarce attention has been paid to the inclusion of ethics contents and courses in food science and technology study programmes at any higher education level aimed to develop an ethical mindset of the graduates and future food professionals. Recently some books on food ethics have been published, whilst clear guidelines, supporting references and documents for training ethics on food aspects are still missing.

Abundant literature can be found on teaching ethics in other subjects, such as medicine and business, but this is scarce in food ethics. This observation prompted the ISEKI_Food-4 network to develop teaching material to promote the implementation of food ethics in the food studies curricula, and this book is a consequence of that will.

This book provides a simple, straightforward framework which can be differently used as a training tool for a specific food ethics discipline/course or as transversal support to curriculum development.

The book consists of a series of chapters divided in three main sections focused on the implementation of the ethical principles in the food value chain and the best practices applied in the food sector.

In the first part, contributions of experts highlight the main societal issues that affect the food value chain and that justify the need of teaching food ethics and provide contents to implement the teaching of food ethics, including a view on bioethics.

The second part includes chapters where ethical issues of the food production chain are described and that can be used by both teachers to prepare training materials and students to become aware of food ethics issues as future professionals or employees. Specific chapters are dedicated to existing tools to promote the professional integrity of professionals and ethical responsibility of food companies like code of ethics and corporate social responsibility.

The third part includes three case studies on ethical issues along with their critical points for a deeper discussion in a classroom.

Due to the complexity of the food product chain and its value in our society, competences of different areas are required to consider all the various aspects of food ethics. Thus, this book has been written by an interdisciplinary team of experts of different areas, including food technologists, experts of food law, food safety, business and economics that agreed to collaborate in the development of this important tool to improve ethical skills and competences of the current and future food professionals.

Coimbra, Portugal   Rui Costa
Teramo, Italy   Paola Pittia

Contents

Contributors

Vitti Allender Cardiff Metropolitan University, Cardiff School of Sport and Health Sciences, Cardiff, UK

Umit Altuntas Istanbul Technical University, Department of Food Engineering, Istanbul, Turkey

Rui Costa CERNAS, College of Agriculture of the Polytechnic Institute of Coimbra, Bencanta, Coimbra, Portugal

Simon Dawson Cardiff Metropolitan University, Cardiff School of Health Sciences, Cardiff, UK

Athanasia M. Goula Aristotle University of Thessaloniki, Department of Food Science and Technology, Thessaloniki, Greece

Mine Gultekin-Ozguven Istanbul Technical University, Department of Food Engineering, Istanbul, Turkey

Harris N. Lazarides Aristotle University of Thessaloniki, Department of Food Science and Technology, Thessaloniki, Greece

Louise Manning Harper Adams University, Newport, Shropshire, UK

Anna McElhatton University of Malta, Faculty of Health Science, Msida, Malta

Adelino M. Moreira dos Santos Coimbra Health School of the Polytechnic Institute, Coimbra, Portugal

Yasmine Motarjemi Consultant, Rue de la Porcelaine, Nyon, Switzerland

Luis Adriano Oliveira Mechanical Engineering Department, Faculty of Sciences and Technology of the University of Coimbra, Rua Luís Reis Santos, Coimbra, Portugal

Beraat Ozcelik Istanbul Technical University, Department of Food Engineering, Istanbul, Turkey

Paola Pittia Faculty of Bioscience and Technology for Food, Agriculture and Environment, University of Teramo, Teramo, Italy

Judith Schrempf-Stirling Robins School of Business, University of Richmond, Richmond, VA, USA

Merve Yavuz-Duzgun Istanbul Technical University, Department of Food Engineering, Istanbul, Turkey

About the Editors

Rui Costa is a Professor of Food Engineering at the College of Agriculture of the Polytechnic Institute of Coimbra. He has a first degree and a Ph.D. in Food Engineering from the Catholic University of Portugal. His research interests include transport phenomena in food processing, safety and environmental aspects of food processing, the shelf life of foods, and use of new protein sources in foods. He has a long experience, at international level, in quality assurance of higher education, curriculum development in food studies, regulation of food professions, and the improvement of the links between research, education and industry.

Paola Pittia is Professor in Food Science and Technology at the Faculty of Bioscience and Technology for Food, Agriculture and Environment of the University of Teramo. Her research expertise is on food processing and product design as well as on quality and shelf life of foods. She has a long and internationally recognized experience on curriculum development and quality assurance in food studies and innovative teaching materials. She was the coordinator of the Erasmus Thematic Network ISEKI_Food-4 and currently is the President of the ISEKI-Food Association, nonprofit, international organization with activities aimed to strengthen the interactions between education, research and industry.

Part I
Introduction

Chapter 1
Teaching Food Ethics

Rui Costa

1.1 Why Teach Ethics in Food Studies?

1.1.1 Ethics and Food Professionals

We all deal with food as consumers and most of the readers of this book deal with it as professionals.

As consumers, selecting the best food ingredients, preparing a tasty meal, and choosing the right diet to assure our health are not the only thoughts we have on food. If a certain food product being sold in a supermarket has been produced recurring to child labor, and if we know it, will we buy it? If a food product uses genetically modified food materials, will all the consumers buy it? If we have the knowledge that producing a certain food pollutes the planet comparing to other food alternatives, will we continue to consume it? The public interest that in the past focused primarily on nutritional aspects of foods includes now ethical aspects of food production including fair trade, novel foods, animal welfare, climate change, and the sustainability of the natural resources (Wharton 2011). Also, as technology continues to change our daily life with general acceptance when it refers to a new communication tool or transport means, when applied to food, not all advancements are felt by consumers as good. Food is a subject where the society is more conservative, as the texts of Michael Pollan testify (Pollan 2008), for which food scandals, thoroughly reported through the media, highly contribute to sow consumer distrust in the food chain.

As professionals working with food, many doubts can come to our mind about issues from the agricultural production, through its transformation until its commercialization to the consumer, from restaurants to highly automatized

R. Costa (✉)
CERNAS, College of Agriculture of the Polytechnic Institute of Coimbra,
Bencanta, Coimbra, Portugal
e-mail: ruicosta@esac.pt

© Springer International Publishing AG 2018
R. Costa, P. Pittia (eds.), *Food Ethics Education*, Integrating Food Science and
Engineering Knowledge Into the Food Chain 13,
https://doi.org/10.1007/978-3-319-64738-8_1

food industries. As professionals, if our company has soft food safety controls, will we have a compliant position? If one's company can buy cheap ingredients from a company with a weak environmental performance, should we buy it? What about if our business supplier uses child labor, would we buy its products? Several ethical labels, such as the Fairtrade and Certified Sustainable Palm Oil, have been adopted by the industry to answer the growing concern of the public about unsustainable industry practices.

Both as consumers and as professionals, a number of questions appear in our daily life that requires us a judgment if it is right or wrong – an ethical judgment. As consumers, the choice of foods affects only ourselves or the family members for which we are responsible, a limited number of people. As food professionals, our actions can affect millions of people. Although some actions are subject to law, that is, continually becoming more rigorous to assure consumer protection and public health, not all actions of the food professionals are scrutinized by law or are able to assure consumer protection or public health, and the professional is sometimes faced with dubious ethical procedures that require him/her to take full responsibility. New moral problems can be listed at any moment as new challenges appear with social and technological changes that require the food professional to make new decisions. Food professionals must then be prepared to reflect ethically before taking their decisions at the different steps of the food supply chain, either in food production, processing, or distribution, and must also be able to engage in the dialogue about food policy.

This makes clear that educational institutions, as universities or polytechnics, offering food studies (food engineering, food science and technology, food chemistry, nutrition, etc.), must prepare their students not only with the technological skills dealing with food and with other employability skills related to communication, group work, etc. but also with capabilities of making judgments on their own actions and taking full responsibility for it, sometimes defending themselves from wrongdoing by taking adequate precautions.

Food and other professionals can find three levels of moral problems: (i) simple wrongdoing in which the solution is evident, (ii) those that are the object of legislation, and (iii) moral dilemmas.

Cleaning well the food processing equipment is a simple moral issue because it is evident that one should prepare foods without any harmful substances to the consumer (and, of course, with the highest quality possible). This problem is also subject to legislation because dirt cannot be found in foods (maximum content of bacteria, etc.). Most of the wrongdoing is evident, with no need to reflect with the help of ethics theory.

However, moral dilemmas can be found in our daily work. An example of a small one: when frying potato, the filters of continuous fryers separate slices of burned potato together with deteriorated oil. That oil may contain more than 25% of polar compounds, a regulatory limit to use of frying oils. However, if the chunks of potato slices are filtered from this oil, the oil can be remixed with fresh oil. The frying producer may deliberately contaminate the oil, though it can make it in a proportion that is able to keep the level of polar compounds below lower that regulatory limit of 25%. Is this a moral issue? Yes. Some professionals are against this practice because the producer intentionally contaminates frying oil. Others are in favor because the legislation is respected – the harm to the consumer does not exist because the law is respected, respecting the oil's limit of polar compounds content – and at the same time a bigger profit and less pollution with residues are obtained.

An example of a complex dilemma is the genetically modified foods issue, a controversial topic that generates very different positions by policy makers in the various countries (the USA vs. EU), with notable differences between the consumers (more critical) and scientists (more favorable). The controversy about this topic does not look to have an end, and even the acceptance by the consumers has been changing along the time. Recently, in the USA, consumers have been increasingly rejecting GMOs even though US regulations facilitate its commercialization comparing to EU regulations.

One should have a clear understanding that an ethical decision is needed when dealing with moral dilemmas, which are issues with different, and quite often, contradictory solutions, not addressed by legislation. However, in this book, the ethical issues presented and discussed include problems that may be already covered by legislation and issues that would lead a philosopher to exclude it as ethical dilemmas because what is wrong is already clarified by law. An argument in favor of debating the ethics of these controversial issues is that even when "solved" by legislative means (e.g., consider a nonfood issue like abortion), the ethical dilemma still continues to exist, still being subject of debate, and that later may be subject to new legislation. To solve controversial and thus complicated issues, the newcomer needs guidelines on ethics to address these problems (Weston 1997). In this book, ethics issues are identified and detailed to make aware graduates of their possible future professional dilemmas and use them to train on ethical analysis necessary for decision-making on new ethical issues.

1.1.2   The Teaching of Ethics in Other Subjects

Food ethics teaching is a relatively new discipline when compared to teaching applied ethics on other subjects. Simple searches on the web and scientific databases, in English, with the search terms indicated in Table 1.1, give a quick idea of the extent of the work worldwide on food ethics and on teaching food ethics compared with analogs of other subjects. The number of hits for "ethics" is 175 million, 10 million for "business ethics," and 6.6 million to "bioethics." "Food ethics" appears at least 50 times less than these discipline examples. More important are the results of the search of scientific texts where similar proportions are observed. These simple, and not rigorous, numbers illustrate that food ethics and teaching food ethics are in its infancy when compared with the teaching of bioethics and of business ethics.

Table 1.1 Number of hits from web and scientific database searches of search terms on teaching ethics in March 2017

Search terms	Web search hits	Scientific texts database (www.b-on.pt)	Scientific texts database (www.b-on.pt): the addition of the term "teaching"
Food ethics	136,000	612	0
Bioethics	6,560,000	16,229	270
Business ethics	9,990,000	14,095	596
Ethics	175,000,000	179,222	3506

This fact can be justified by two main reasons. One is the fact that food professionals, at the industry level, do not deal directly with the consumer, while medical doctors deal directly with the patient. Consequently, the policies of assuring consumer protection and public health have been focusing on companies and not on professionals. This fact did not speed up the need for regulation of food professions and thus the adoption of deontological codes, which consequently pulls the development and teaching of ethics in the respective subject area. This same fact instead pushed the development of corporate and business ethics. Another justification is the recent development of food science and technology (Kostaropoulos 2012), only from the 1940s, that justifies the smaller number of food professionals when compared to medicine, business, or engineering, for example.

Another evidence of this short history of teaching food ethics is the appearance of a scientific journal on *Food Ethics* (published by Springer), only recently, in 2016, while, for example, the *Journal of Business Ethics* (also from Springer) started more than 20 years before, in the year of 1982.

The lack of bibliographic references on teaching food ethics justifies the use in this chapter of references from other subject areas, particularly from business and bioethics.

1.1.3   Notes on the History of Ethics and Teaching Ethics

To better understand the difference of teaching ethics in the different subject areas, a short overview of history helps.

The birth of ethics is usually wrongly attributed to the Greeks, at 530 BC by Pythagoras of Samos (Bourke 2008). Ethics related to professionals, specifically in medicine, was found to date back at least 1750 years BC (Kuhse and Singer 2009), with the Babylonian Code of Hammurabi stating obligations of the doctors to the patients. Later, in the health professions, nursing ethics started during the nineteenth century, very much related to the official appearance of this profession. Contemporary bioethics began to take shape in 1960 (Kuhse and Singer 2009).

According to Zwart (2000), food ethics is "as old as morality itself." The Greeks were worried about the virtue of temperance, and the Jewish made distinctions between legitimate and illicit foods. Ethics related to the consumer has a mark in history with the boycott of slave-produced sugar in 1791 in the UK, but some attribute the first consumer ethics example to Pythagoras – he was thought to be vegetarian (Newholm et al. 2015). Nowadays, apart from the food issues reported with detail along this book, what will have a spotlight in history is the global hunger and the human right to food, stated in the Article 25 of the Universal Declaration of Human Rights (1948) and the Rome Declaration on World Food Security (1996) (FAO 2001). These statements claim that hunger is the result of insufficient effort to guarantee the universal right to food.

The start of the teaching of ethics cannot be distinguished from the teaching of history and philosophy. However, the teaching of ethics for the professional is connected to instruction for these professions, particularly when the related profession is regulated, as the oldest example of medical ethics illustrates. Davis (1999) observed a boom of interest in ethics since the 1970s primarily related to professional ethics and other forms of applied ethics. The evidence is the number of corporate codes of ethics that flourished in companies, the codes of ethics of professions/occupations, the number of courses on ethics in curricula, and the ethics committees and ethics consultants that, in health, are at least found since the 1960s (Moreno 2009). However, this boom did not affect every single subject. Ethics teaching in veterinary degree programs should be present since the veterinary profession is regulated in many countries, but a search in 85 European faculties out of the 99 available at the Federation of Veterinarians of Europe website revealed that this is not the case (Magalhães-Sant'Ana et al. 2010).

The option for implementation of an ethics course in a specific degree program is dependent on the status of the profession(s) associated with the degree program. Disciplines with a long past of teaching ethics, like medicine or law, usually include an ethics course in the program degree curriculum that covers philosophy basics, current neuroscience knowledge of self-consciousness, and the application of ethics issues in the profession (see Chap. 2). This course is usually required in degree programs that give access to regulated professions which must apply a deontological code. The relevance of ethics for the profession is not important for many actors in the education of food science and technology subject due to the almost nonexisting regulation of related professions in Europe and elsewhere (Costa et al. 2014). On the other side, the lack of regulation of food science and technology professions is explained by the relatively new history of this discipline, with less than 80 years (Kostaropoulos 2012).

1.1.4 Driving Forces to Teach Ethics in Food Studies

The driving forces to teach ethics come from education standards, deontological codes of professional associations, and a general approach of development of transferable skills in current education trends (Illingworth 2004).

Education standards for food studies have been only recently developed. The European Erasmus networks FOODNET in 2000 and ISEKI4 in 2013 concluded that ethics was a discipline not usually present in European program degrees at any

Table 1.2 Reference to ethics in international education standards of food science and technology

Association/agency	Group of learning outcomes	Learning outcome
Quality Assurance Agency (UK)	Self-management and professional development skills	Recognize the existence of moral and ethical issues associated with the subject (threshold level)
Institute of Food Technologists (USA)	Generic skills – professionalism skills	Commit to the highest standards of professional integrity and ethical values
ISEKI Food Association (EU)	Generic competences – communication abilities, ethics, and personal skills	(Without indication of a learning outcome)

level (bachelor, master, or Ph.D.). But while many teaching staff may be resistant to include ethics in the curricula, competency on ethics is widely felt as essential to be part of undergraduate food studies programs and was embodied in several international standards, such as the European Quality Assurance Scheme of the ISEKI Food Association (IFA 2016), the North American Institute of Food Technologists education standards (IFT 2016), and the UK Quality Assurance Agency benchmarks (QAA 2009). These standards refer slightly to ethics in teaching food science and technology studies as compared with the references to other sectors, even in the same document (see agriculture in QAA (2009)). As an example of recommended learning outcomes, in the last reference, the threshold performance related to ethical skills, part of "self-management and professional development skills," is to "recognize the existence of moral and ethical issues associated with the subject" (Table 1.2).

Teaching ethics in food studies is proposed for the benefit of professionals, consumers, and the society at large, in particular to the food industry itself (Early 2002). The panorama of teaching ethics in food studies programs could be entirely different if the food professions were regulated. As this is hardly a common situation in all continents (Costa et al. 2014), only a few examples of deontological codes for food professionals can be found (see Chap. 6).

In Europe, since the Bologna reform in higher education, more attention has been given to employability skills, part of them also classified as generic, transferable, or soft skills, i.e., skills not dependent on the subject, such as communication and group work skills, the two most cited. Ethics skills have also been included among the generic skills that need to be developed in the learner but not with the importance as the two referred before (Flynn et al. 2013). In the standards referred before (QAA, IFT, IFA), ethics skills are exactly inserted in the group of generic skills and due to this fact have deserved more attention.

1.1.5 What Is Ethics?

This chapter started with the need of teaching food ethics and the citation of other subject areas ethics texts. However, to go on further, it must be clear to the reader what is ethics and its related terminology.

For philosophers, ethics is the philosophy of moral values, the study, and reflection of what is wrong and what is right. Since it is related to moral, ethics is often perceived as vague, and the interchangeable use of both terms contributes a lot to that perception. It is common to listen to the media and may be within circles of contacts of the reader, the accusation of "lack of ethics" by someone, usually with the meaning of the lack of basic moral values. Clark and Ritson (2013), in its Chap. 3, clarify well the correct use of the terms moral and ethics: moral is more used to refer to society's standards of acceptable behavior, whereas ethics relates to the systematic approach to morality.

Ethics can be classified into three levels: metacthics, normative ethics, and applied ethics. Metaethics is related to the fundaments and meaning of ethical principles. Ethical principles are also called normative ethics. These guide our judgment to what is right or wrong. Applied ethics is the use of normative ethics to decide on issues, simple or controversial ones, like abortion, animal rights, making use of genetically modified organisms, etc. Within applied ethics, professional ethics is of particular interest to this book.

There are many ethical principles. Probably the most well-known ethical principle is what is designated as the Golden Rule: "don't do unto others what you don't want others to do unto you." However, no single rule can answer all dilemmas. The main principles are here highlighted. Virtue principles place emphasis on developing good habits of character, such as justice. Duty theories are based on the feeling that there are clear obligations we have as human beings, such as not harm others. These theories are designated as deontological. Consequentialist theories are based on cost-benefit analysis of an action's consequences, either to one's, to others, or all. If the consequences are more favorable than unfavorable to everyone, this is called utilitarianism.

The difficulty in using normative ethics is that between the hundreds of these norms, there are many contradictions or some lead to contradictory results. To solve an applied ethical issue, different normative principles are used, depending on who uses it. These principles should be considered as valuable and valid by people on both sides. The literature on normative ethics is abundant, and its general knowledge is familiar to any higher education student, who learned it during secondary school. Normative ethics will need to be recalled in higher education, either in a lecture or on homework reading.

Some food issues identified in this book may not look as moral issues because, for example, they are regulated by law. However, if contradictory positions exist around it and if it is a distinct moral issue, the conditions to be considered an ethics issue are fulfilled (Fieser and Dowden, n.d.).

Professional ethics is implemented through norms that apply to members of a profession (liberal profession or occupation). The norms are partly different from profession to profession due to their different nature but have a common ground since their ultimate practical aims are common, such as that of consumer protection and/or public health. The different technical nature and whether the professional deals directly or not with the end stakeholder – medical doctors deal directly with persons (patients), while engineers prepare equipment to be used by persons (consumers) – are at the root of the differences between the norms of these professions.

1.2   Objectives of Food Ethics Courses

1.2.1   Aim of an Ethics Course

The ultimate aim of an ethics course is to develop moral cognition in the student (Bowden and Smythe 2008). On its turn, moral cognition, the making of ethical choices (Zahn et al. 2011), implies the person to have ethical sensitivity and moral courage.

When converting the aim of a course into course learning objectives, the ideal learning outcome of a food ethics course would thus be the demonstration of moral cognition by the student. However, any type of demonstration of moral cognition can hardly be used as a learning outcome as it is impossible to evaluate it. Only in few life situations, outside of controlled learning environments, could the teacher evaluate the student about this valence. However, even studies done in work environments could not measure it, as referred to a study by The Centre for Vocational Assessment and Research, which study focused on the assessment of attitudes, ethics, and behavior at work (Mossop 1997; referred by Bowden and Smythe (2008).

An additional difficulty is that ethical behavior could only be evaluated consistently by studying students' attitudes over time, but attitudinal surveys would not be adequate, as referred by Enghagen (1990) (cited by (Lynn 2010)). Not only is impossible to assess it, but this complex learning objective may also be out of the teachers' influence. Moral cognition has been known to be affected by the brain characteristics since the crucial observations for the beginning of the modern moral neuroscience, of changes in the moral behavior of individuals after certain brain lesions (Zahn et al. 2011). Thus, teaching ethics does not mean that that is possible to teach anyone to be ethical.

Typically the educational institution can only focus on observable behaviors in the classroom and plan to assure that the student acquires the appropriate knowledge and has practices that assure he/she masters it and can discuss ethics subjects. Teaching ethics is useful if it raises the awareness of ethics, increases the knowledge about ethics topics and ethics practices, and promotes the capabilities of discussion. Several authors claim that teaching ethics can effectively develop students' moral reasoning skills and ethical sensitivity (Bowden and Smythe 2008; Sims 2002; Weber and Glyptis 2000).

It is widely accepted that ethical sensitivity can, in fact, be observed in the classroom and be built in the student. Ethical sensitivity (also called ethical radar by Johnson (2009)) is the recognition and the interpretation of the ethical dimension in the practical situation (Weaver 2007) and is a preceding development needed to achieve moral cognition. The development of ethical sensitivity in the student can be accomplished by covering all ethical practices throughout the curriculum and by developing argumentative capabilities in advocating an ethical position (Bowden and Smythe 2008). Students must formulate well-justified positions based on critical analysis and reflection. Students need to reflect critically on each issue (Kaltoft and Sandøe 2001), balancing its values with others' (colleagues, public, etc.) opinions, and taking into account normative ethics.

Ethical sensitivity is a requirement for the making of ethical choices, but it needs moral courage at the same time to make it reflected positively into practice. Ethical sensitivity is particularly required in organizations because these are composed of individuals with different values that must be respected, for which "a collective agreement on personal integrity is not likely" (Loe and Ferrell 2001). However, one has to be conscious that awareness of ethical issues does not ensure an ethical behavior from the individual, as an ethics trainer for some Fortune 500 companies and the president of the Academy of Management testify (Thompson, 2006; referred by Bowden and Smythe (2008)).

The development of ethical sensitivity is here proposed for teaching professional ethics. Using professional examples in a professional-related class helps to motivate students, making clearer to them the usefulness of ethics, and helps students to be prepared for their professional lives. At the same time, other generic skills, as communication and reflection, are developed (RAE 2011).

1.2.2   Course Outcomes

Ethics is a social-humanistic topic, contrasting on its nature with almost all subjects that are part of a food science and technology curricula that are of technological nature. Naturally, this topic becomes less comfortable both to be taught by the food science and technology specialist teacher and to be learned by the food science and technology student. The content and the objectives of this course must thus be put in an easy way to attract both teachers and students and especially to make the last ones realize how useful it can be for their future profession to be involved in an ethics course and to analyze and discuss ethics topics and case studies.

An overview of ethics courses' objectives taught in different subjects is presented in Table 1.3. The examples cited are written in a variety of ways, either as:

- Objectives for teachers: e.g., "teach students the profession's established code of ethics"
- Objectives for students: e.g., "be able to identify the ethical elements in decisions"
- Some presenting broad objectives: e.g., "develop critical thinking skills and professional judgment"
- Objectives related to a professional act: e.g., "the ability to obtain a valid consent or refusal of treatment"
- Objectives that natural outcomes of an ethics course: e.g., "ensure students recognize and appreciate the unavoidable ambiguity in ethics"

Common objectives between the examples shown can be observed, particularly between the business and human services and counseling objectives. Engineering is more professional oriented, and medicine is highly professional specific. Some objectives may be doubtful to reach, such as "acquiring moral courage" or "improving moral cognition," as explained in the previous section.

To the author's opinion, the simplest, minimum, formulation of objectives can be condensed in two objectives, either for a course on ethics or integrated into the curriculum approach, which assures the development of ethical sensitivity by the student. Course objectives will be put here as learning outcomes. These are "statements of what a learner knows, understands and is able to do on completion of a learning process, which are defined in terms of knowledge, skills and competence" (EU 2008).

Table 1.3 Example of ethics courses' objectives in different subjects

Business Brinkmann and Sims (2001)	Human services and counseling Corey et al. (2005)	Engineering (RAE 2011)	Medicine (Goldie 2000)	Training in companies Webber (2007)
Know thyself, your own moral values, and thresholds	Improve students' self-knowledge	Be able to identify the ethical elements in decisions	The ability to identify the moral aspects of medical practice	To engender trust and confidence among stakeholders
Learning to see moral issues, conflicts, and responsibilities	Improve ethical sensitivity	Develop critical thinking skills and professional judgment	The ability to obtain a valid consent or refusal of treatment	To stimulate mutual moral development through self-discovery in peer group discussions of real ethical dilemmas
Acquiring moral courage	Improve moral cognition	Understand practical difficulties of bringing about change	The knowledge of how to proceed if a patient is only partially competent or fully incompetent	To develop a concrete plan for a career-long, ongoing moral enhancement of individual attitudes, and planned business behavior
Learning how to handle moral issues and conflicts	Instill determination to act ethically	Develop a professional ethical identity to carry forward in their working life	The knowledge of how to proceed if a patient refuses treatment	
Acquiring a critical attitude toward the business school curriculum and its disciplines	Teach questioning of the ethical dimensions of their workplace	Be able to address and resolve problems arising from questionable practice	The ability to decide when it is morally justifiable to withhold information from a patient	

(continued)

Table 1.3 (continued)

Business Brinkmann and Sims (2001)	Human services and counseling Corey et al. (2005)	Engineering (RAE 2011)	Medicine (Goldie 2000)	Training in companies Webber (2007)
Learning to identify the specific moral aspects of a situation	Ensure students recognize and appreciate the unavoidable ambiguity in ethics, i.e., of multiple points of view or contradictory possibilities	Understand the nature of professional responsibility	The ability to decide when it is morally justified to breach confidentiality	
Learning to share moral understanding	Instill in students the idea that there are multiple pathways to addressing a single ethical dilemma		The knowledge of the moral aspects of caring for a patient whose prognosis is poor	
	Teach students the profession's established code of ethics			
	Teach students their legal, ethical, and professional responsibilities			

The two learning outcomes can be:

1. The learner must demonstrate knowledge about food-related ethics topics and ethics practices.
2. The learner must demonstrate argumentative proficiency in dialogues about food ethics issues.

The first learning outcome is a basic one as the student only needs to identify topics of ethics and examples of the food science and technology area where ethics is relevant for decision-making. Knowing them proves that the student is aware that not all acts related to his/her profession are the simple execution of technical guidelines but that sometimes a new decision must be taken for which guidelines are missing and are contradictory or simply that the decision requires moral reflection absent from standards or guidelines.

The second learning outcome builds on the achievement of the first and goes beyond it. Being capable of argument in ethical issues gives students a real tool to face delicate professional situations and more confidence on the reflection that he/

she as a professional must undergo before making a decision. Having trust in ethical reasoning and/or strong moral values facilitates the manifestation of moral courage which is the ultimate aim of an ethics course. Argumentation is also more efficient in making the student realize the need for the incorporation of ethics topics in the curriculum and thus more self-motivating to the student.

Practicing argumentation, in groups during classes, is an active learning activity that effectively makes the student aware of the need for reflection on the dilemma and of his/her weaknesses or limited self-knowledge, which could be either related to insufficient technical knowledge that impedes prompt answers or solutions and/or insufficient reflection on ethical principles. If argumentation activities (debates, dialogues) cover a broad range of ethical issues of the food chain, the achievement of the first learning outcome is also assured. The second learning outcome also requires communication skills and capability of persuasion. These skills may also be identified and detailed as a learning outcome, according to the particular aims of the course.

1.3 Implementing a Course on Food Ethics

1.3.1 Before the Implementation

The teaching of ethics must prepare the future professionals to tackle expected problems of their working lives. This can be achieved by tailoring the learning and teaching environment in such a way that students understand that this topic is as relevant to their subject area as technical themes and that themselves should deal with it comfortably and expect to be able to achieve a reasonable level of expertise (Illingworth 2004).

Since the objective of teaching ethics is not to transfer a set of well-defined scientific knowledge but to promote change or improvement in attitude and awareness and to develop capabilities for ethical analysis, several possibilities may be adequate, mirroring ethics itself where it is common to find several possible solutions for a single problem. Most naturally, all the resources needed to teach ethics are possible to find in every higher educational institution. Each program coordination team needs to plan in which course(s) and with each teacher whether they can implement the teaching of ethics, as it is usually done with other generic skills, such as communication and group work (Davis 1999).

As referred before, few references on teaching food ethics can be found. Although a few percentage of food studies curricula include an explicit focus on ethics, many teachers of those programs agree that the topic must be taught. However, those wanting to teach it will pose themselves some questions. Who should teach it? Which topics should the curriculum include? How to teach it?

1.3.2   Philosopher or Science and Technology Teacher?

The question above is the first question that is raised when the idea of teaching ethics is discussed. Having a philosopher as a teacher of a food ethics course is, in practice, difficult due to the absence of this specialty in many faculties or colleges of food science and technology and also due to the general feeling of science and technology teaching staff that, with a philosopher, the ethics course would be expected to be less practical, contrasting with the common practical teaching approach in food science and technology. In fact, the pioneers of medical ethics were moral philosophy teachers and lectures were the main teaching method (Goldie 2000). A comparison between these two kinds of teachers was made by Downie and Clarkeburn (2005), who compares a philosopher teaching ethics to science students to a statistician teaching statistics to biology students. The authors claim that a biologist with the practice of teaching statistics is better prepared to teach statistics to biology students than a statistician.

The food science and technology teacher must, of course, undergo training to teach ethics, which is anything but impossible. For a higher education teacher, it is common to master new knowledge during their research and teaching experiences, and ethics can be just one new topic to master, as long as the teacher is enthusiastic to tackle this new subject.

For the food science and technology teachers, ethics is for sure no new subject in their social and professional lives. What is required to teach ethics is to present some more information and to be prepared to interact with students on this subject.

In those institutions with philosophers in their teaching staff, a collaboration between a philosopher and a food science and technology teacher would be, for sure, the most enriching experience, for the faculty and the students.

Whatever the main teacher specialty, one must have the notion that students want guidance from teachers, in technical aspects, but they may also do it in ethical reasoning and moral values, an influence that is not closed to the classroom (Lynn 2010). The teacher will thus face the doubt of providing its judgment on ethical issues particularly because promoting students' critical skills should always be present in the teacher's mind (Loe and Ferrell 2001).

1.3.3   A Single Course or Across the Curriculum?

A separate course in ethics is often used, from 15 weeks duration to intensive 3 weeks (Kaltoft and Sandøe 2001). However, this is not always acknowledged as being the best option. Layman (1996, reported by (Goldie 2000)) identified four curricular arrangements in medical ethics teaching: (1) integrated modules across the curriculum, (2) a single discrete course and integrated modules across the curriculum, (3) multiple courses, and (4) multiple courses or seminars and clinical rotations.

In medicine, there is a broad consensus that teaching of ethics should be integrated horizontally and vertically in the medical curriculum, including within professional settings, to make clear to the student that ethics is not an isolated subject (Goldie 2000). Moreover, if ethical education is promoted in a professional setting, it is seen as even more efficient.

From the examples from other subjects, for a relatively recent discipline as food science and technology, an integrated approach to the curriculum would be highly recommended. This integrated approach requires coordination between teachers or at least a plan of the distribution of topics between several disciplines. In this case, an ethics teaching coordinator is needed to help the other teachers to implement the discussion of an ethics issue. This strategy might be difficult to implement if most of the teachers do not feel competent to approach this issue in his/her traditional teaching subject.

A strategy to teach ethics across the curriculum can be made starting at an introductory ethics level at first year to higher demanding professional ethics in the last year, as can be found for medicine (Peer and Schlabach 2010), practice administration (Schwartz 2009), and engineering (RAE 2011).

Bowden and Smythe (2008) suggest that chosen disciplines should cover ethical issues related to those disciplines (e.g., GMOs in genetics) and additionally a basic ethics content related to modules where ethics is taught, such as ethical theory, public interest disclosures, and code of ethics. This approach enables students to perceive the relevance of ethics, and at the same time, there is no disruption during teaching provision (RAE 2011).

RAE (2011) recommends to prepare the introduction of ethics into existing curricula following three steps:

1. Analyze in which modules/courses ethics analysis and case studies are already present, though it may not be clear to every teacher of the program degree, particularly because it may not be titled as ethics.
2. Identify which other modules/courses can discuss ethics naturally (e.g., professional modules).
3. Plan the implementation with the respective teachers of the modules/courses and provide them support, if needed.
4. Include a gradual level progression along the curriculum.

In food ethics, an across-the-curriculum approach could be planned based on the topics and case studies described in Part II to Part IV of this book, being discussed in related disciplines, for example, ethics of consumption in a marketing discipline and food industry issues in a food processing discipline.

An example of a quite original format is an integrated online course that lasts as long as the program degree lasts. On a hotel and restaurant management program degree, faculty of core courses reminds students to complete the module tasks including assignments, but no teacher, except for assignment evaluation, is involved (Lynn 2010).

A short paragraph on ethics training in the industry. Companies need to promote this training if employees demonstrate short ethical skills, particularly on corporate social responsibility. Training in the industry is suggested to be implemented with a strong management commitment, taught by a professional trainer, of at least 4 h duration and preferably starting with new employees, with a follow-up 2–6 months later (Bowden and Smythe 2008).

1.3.4   Teaching Content

Ethics, by definition, means moral philosophy. This implies that to discuss ethics, the content of instruction must include philosophical basics of ethics as utilitarian ethics, virtue ethics, etc. Depending on the fact of how fresh is this knowledge from previous education, especially for an integrated into the curriculum approach, a review of these theories can be planned as a preliminary student reading and/or with a lecture.

Presentation and discussion of food ethics issues (social and professional) should be the next content and the core of the food ethics course. These should include all possible professional situations that require an ethics analysis and grab the student to the topic.

One or more applied ethics tools like the ethical matrix can also be used by the teacher to be implemented in class exercises. These give the ethics problem-solver a framework that helps him/her to study the issue, reflect on it, and reach a decision.

1.3.4.1   Ethics Theories

As referred in Sect. 1.1, there are many ethics theories, but the main ones are the virtue, the duty, and the consequentialist theories.

Virtue principles are those referring to uncontested rules recognized as virtues such as courage, justice, generosity, and sincerity. Following these rules, considered as good habits of character, implies not following contradictory rules such as cowardice, injustice, insensibility, and vanity.

Duty theories are based on what many feel as obligations that a human must have with others, such as not killing. There are several subdivisions of theses duties, from which one of them is here presented and which titles are self-explanatory: duties of beneficence and non-maleficence, duties of special care, duties of honesty and fidelity, duties deriving from agents' and patients' histories of conduct, duties of reciprocity and fair play, further duties of justice, and duties to other species (Darwall 2005). These theories are also called deontological, a word derived from the Greek *deon* (duty).

The consequence of our actions is certainly one of the most common reflections we do before deciding on an action and is called consequentialism. If we judge the result from that action more positive than negative to everyone involved, this is called utilitarianism. Another form of consequentialism is the hedonist value theory, according to which pleasure is the only intrinsic good (Darwall 2005).

Of course, all these theories, when considered simultaneously to make a decision, can lead to conflicting decisions. Recalling these theories in the classroom or as an assignment will help students to be prepared for the discussion of case studies.

1.3.4.2 Food Issues and Case Studies

The ever-increasing demand for a close relationship between university and business justifies, by itself, to teach ethics around the food topics. Graduates must be ready to quickly integrate companies and perform well a professional role, which may demand an adequate answer to less ethical practices within the first job.

The specific content suggested for ethics for food science and technology is presented in this book:

   (i) Sustainability along the food supply chain (Chap. 3)
  (ii) Ethics of consumption (Chap. 4)
 (iii) Food industry issues (Chap. 5) and case studies (Chaps. 11, 12, and 13)
  (iv) Codes of ethics in food professions (Chap. 6)
   (v) Corporate social responsibility (Chap. 7)
  (vi) Whistle-blowing (Chap. 8)
 (vii) Food safety risk communication (Chap. 9)
(viii) Publication ethics (scientific writing) (Chap. 10)

These issues are the main professional issues that a food professional must be aware of. They provide an insight into the profession and guidance to professional practice. Other narrow and recent topics, such as nanotechnology, for example, can be found elsewhere in the available literature.

New case studies ready to be used in the classroom are presented in Chaps. 11, 12, and 13. For case studies, guest speakers from industry, research, and teaching can also be invited, bringing an extra enthusiasm in discussing ethical dilemmas. As in other matters, having other speakers than their well-known teachers helps students to open their minds to these problems, especially those that deal with these issues in their daily professional life.

1.3.4.3 Ethical Decision-Making

Maybe the most straightforward and efficient way to attract students to ethics is to use it in problem-solving and decision-making on professional problem case studies. Bloisi et al. (2006) provide a rational to problem-solving, starting with problem awareness and problem definition, followed by decision-making, action plan implementation, and a follow-through step, and suggest to do it ethically, adding ethical reflection in the decision-making step.

Several frameworks have been developed for the food and other science and technology subjects. The ethical matrix, proposed by Ben Mepham in 1996, is suggested by the British Food Ethics Council and is the best-known framework developed for food issues. It consists of reflecting on three principles applied to the different stakeholders relevant to the dilemma (e.g., producers, consumers). The three principles incorporate the main ethical theories: (1) respect for well-being represents the major utilitarian principle; (2) respect for autonomy represents the major deontological principle; and (3) respect for fairness that is important to both the utilitarian and deontological principles also incorporates the social contract, the view that one's moral obligations are dependent on an agreement among them to form the society in which we live.

Other ethical decision-making models can be found in literature and are worthwhile to explore, such as the moral maps from the Global Food Ethics project (Berman Institute of Bioethics 2015), the value chain analysis (Shani et al. 2013), or the engineering problem-solving using techniques titled as line drawing or flowcharting from Fleddermann (2012). Experienced in the teaching of business ethics, Bowden and Smythe (2008) suggest several nonsubject-specific frameworks of norms.

1.3.5 Teaching Methods

A question may be raised by some readers, whether the environment for ethics learning should be the classroom or a workplace environment. For contrast, let's consider medicine, where it has been pointed out that ethics teaching should be where the action is (Goldie 2000). Medical doctors deal directly with patients, almost the entire time during the working schedule, and which relations may provoke many situations where ethics may be pondered.

This is not the case with food science and technology professionals, who spend a significant portion of their working time with materials and equipment and less with people and their social interactions are much more collaborative than the doctor-patient relation that is more hierarchical, with the actions of the doctor on the patient having consequences on the last one's life. These facts may justify a less demanding professional environment on teaching food ethics. Another difference is that food science and technology professionals take many decisions in groups, in contrast to the individual responsibility of a medical doctor. This particular characteristic requires that most of the decision-making must be done in a group, after reflection and discussion in groups.

In the classroom, teaching methods are common between subject areas, whether in business ethics, hospitality ethics, or medicine ethics. Lectures are commonly used to present theories and ethical decision-making frameworks followed by experiential-based learning techniques, for which there is substantial evidence of its appropriateness to teach ethics. Small group (four to five) class discussion and case analysis are the most effective activities in teaching food ethics (Loe and Ferrell 2001; Goldie 2000; Lynn 2010). These strategies are reinforced by studies about the effectiveness of ethics teaching.

Not explored here, essay writing is seen as a useful learning technique in ethics education (Lynn 2010) though is usually considered as a difficult task to science students (Johnson 2009).

A few teaching methods, commonly used in the teaching of ethics, are presented below.

1.3.5.1   Case Studies

The reflection on real problems, either societal or professional, motivates and engages students in ethics. Societal dilemmas are the simplest to recall and gather the attention of students because most probably they already reflected on it. Nowadays examples, and having in mind that are significant differences between what is legal and the cultural acceptance between countries, are probably still the legality of abortion and euthanasia. More food professional related, but definitely also a societal issue, are the genetically modified organisms (GMOs) used as food. The teacher can make use of any of these issues to engage students in an ethics debate without any special preparation by the students.

Following a student-centered approach, the teacher should plan a sequence of activities and topics starting with what the students know more and is more aware of and then direct them toward the knowledge and skills needed to be able to develop the ethics skills. According to Børsen (2005), in line with this thought, teaching ethics to engineering students is more efficient starting by societal problems and ending on ethical value systems.

Working in small groups on real problems is known to be the most efficient technique in many disciplines due to an active learning attitude promoted in the student, but this is particularly true on ethics teaching (Doorn and Kroesen 2013), where argumentative capability and communication skills are necessary and can only be developed with others. The case studies can be used in different ways, either using presentations, debates, or dialogues, mainly: (i) an introduction followed by a small group discussion and then reporting to the entire class the conclusions of each group, (ii) an introduction followed by group work and a presentation to the whole class some days or weeks after, and (iii) an introduction followed by group work and debate using role-play.

1.3.5.2   Role-Playing

On each case study, possible conflicting positions can be identified. For example, in the GMO issue, the scientist values the knowledge of genetics, while the conservative consumer acknowledges the natural food, without processing and closing to the farm origin, etc. A debate with students having the role of representing the controversial positions of several of these different stakeholders can be easily assembled.

Role-playing in the debate of case studies, which requires students to reach ethical decisions during the interaction with the class, is a highly effective learning activity (Kroesen and van der Zwaag 2009). Students should play roles that need to have argument positions that aren't initially their own, probably outlining thoughts to defend values that aren't their own and involving in a live debate that is emotionally capturing making the students more conscious on other person attitudes. This develops the students' ethical radar and grows in the student a more tolerant attitude toward differences of opinion. Enabling the student to view and appreciate the complexity of ethics, it is highly effective in developing critical analysis and argumentative skills.

Ideally, the role-play scenario should have a few typical characteristics (Loui and Gunsalus 2006): be familiar and contain stereotypes to enable students to be comfortable with their roles, have enough information to be engaging and challenging, and contain an evident dilemma.

1.3.5.3 Socratic Seminar

Some discussions on ethics are referred to as Socratic seminars. Socrates statement "the unexamined life is not worth living" was the motto for these debates that are more useful if done as dialogues. A Socratic seminar (or dialogue) starts with an ethical question that can be answered only by thought (e.g., no need of material testing), as, for example, "should the food industry stop producing caloric foods (as the fat and sugar rich products)?" and continues with more questions based on the answers, e.g., if a reply is "yes, because the average consumer cannot decide on the adequate food intake," and then another question may be "what is a correct food intake?," etc. The critical reflection, based on questions along the dialogue, continues until the original question is exhaustively explored. A final answer may be reached or not. According to Littig (2004), when using this method, judgments on the opinions should not be accepted, and a balance between taking a position and resigning should be pursued. The dialogue should be seen as cooperative critical thinking exercise to test the validity of each one's opinions and previous thoughts on the topic of the question.

1.3.5.4 Tools

Texts describing ethics issues, either books, papers, or newspaper articles are the most abundant materials, though nowadays, media pieces, such as video and podcasts, are also easily accessible (Wharton 2011; Loe and Ferrell 2001). Either the description of the ethics theories, the ethics problems, or the position of the different stakeholders over the various ethics issues can be illustrated with any of these tools.

Web-based software has also been developed, helping the student to practice, individually, ethical understanding and moral reflection: see Agora (www.ethicsandtechnology.com) and the Food Ethics Dilemma (www.foodethicsdilemma.net). These are tools that intend to enable an insight on the ethics issues without the pressure of having to the argument against a colleague or being evaluated. Thus, it is helpful for the student that wants to reflect at its pace, probably before going into a role-play or class presentation.

Less common tools can also be found. For example, the use of games in the business ethics teaching (Loe and Ferrell 2001). These are considered to promote the importance of ethics and comprehension of corporate policies.

1.3.6   Evaluation

Evaluation of an ethics course learning outcomes is a difficult task. This is observed even in subjects where the teaching of ethics has far roots in the history, as is the case of medical ethics (Goldie 2000).

Summative evaluation for grade purposes can be done on the evaluation of debates and of oral presentations, evaluating the range of knowledge, either technical specific of the problem and of moral philosophy or grounded argumentative capabilities shown by the student. Evaluation of written tests enables the assessment of written communication skills but lacks the real-life characteristic of human interaction in ethical issues dialogues.

Measures of moral development have also been used for evaluation purposes (Goldie 2000). Kohlberg, in 1976 (Kohlberg 1976), formulated a "cognitive-developmental theory of moralization" still used nowadays, which divides moral development into three groups of two stages each. The pre-conventional stages (1 and 2), when an individual-concrete perspective prevails, are characteristics of children under 9 years old and other less moral developed persons such as adult criminals. The conventional stages (3 and 4) are characterized by a society-member perspective, conforming to social rules without critical thinking, and are the stages reached by most adolescents and adults. The post-conventional stages (5 and 6), usually reached by adults older than 20 years old, are characterized by one's formulation and adoption of abstract concepts (e.g., the equality of human rights) rather than the adoption of real rules of behavior. The ethics teacher should be aware of theoretical moral stages in order to better understand the students, plan the teaching, guide students, and understand probable differences in students' moral development.

According to Noel and Hathorn (2014), a few variables influence ethical behavior. Apart from age, as referred in the last paragraph, some studies report that women demonstrate higher cognitive moral development, and some personality traits are also correlated to higher development, such as openness to experience and conscientiousness.

Whatever the assessment criteria chose, the alignment of learning outcomes, learning activities, and assessment must be a rule for the effectiveness of ethics teaching, as is recommended for the instruction of any subject (Biggs 2007).

References

Berman Institute of Bioethics (2015) Moral maps. http://www.bioethicsinstitute.org/wp-content/uploads/2015/05/Moral-Maps_Final-web-version.pdf. Accessed 8 Mar 2017

Biggs J (2007) Teaching for quality learning at university third edition teaching for quality learning at university. High Educ 9:165–203. doi:10.1016/j.ctcp.2007.09.003

Bloisi W, Cook C, Hunsaker PL (2006) Ethical problem-solving and decision-making. In: Management and organisational behaviour. McGraw-Hill Education, Maidenhead, pp 476–522

Børsen T (2005) Teaching ethics to science and engineering students. In: Rep. from a Follow. Symp. to 1999 World Conf. Sci. Copenhagen, April 15–16, 2005. http://portal.unesco.org/shs/en/files/8735/11289332261TeachingEthics_CopenhagenReport.pdf/TeachingEthics_CopenhagenReport.pdf. Accessed 8 Mar 2017

Bowden P, Smythe V (2008) Theories on teaching & training in ethics. Electron J Bus Ethics Org Stud 13:19–26

Bourke VJ (2008) History of ethics: Graeco-Roman to early modern ethics. Axios Press, Edinburg

Brinkmann J, Sims R (2001) Stakeholder-sensitive business ethics teaching. Teach Bus Ethics 5:171–193. doi:10.1023/A:1011461418842

Clark JP, Ritson C (2013) In: Peter Clark J, Ritson C (eds) Practical ethics for food professionals: ethics in research, education and the workplace. Wiley, Hoboken

Costa R, Mozina SS, Pittia P (2014) The regulation of food science and technology professions in Europe. Int J Food Stud 3:125–135. doi:10.7455/ijfs/3.1.2014.a10

Corey G, Corey MS, Callanan P (2005) An approach to teaching ethics courses in human services and counseling. Couns Values 49:193–207. doi:10.1002/j.2161-007X.2005.tb01022.x

Darwall SL (2005) Theories of ethics. In: A companion to applied ethics. Blackwell Publishing Ltd, Oxford, pp 17–37

Davis M (1999) Ethics and the university. Routledge, London

Doorn N, Kroesen JO (2013) Using and developing role plays in teaching aimed at preparing for social responsibility. Sci Eng Ethics 19:1513–1527. doi:10.1007/s11948-011-9335-6

Downie R, Clarkeburn H (2005) Approaches to the teaching of bioethics and professional ethics in undergraduate courses. Teach High Educ. doi:10.3108/beej.2005.05000003

Early R (2002) Food ethics: a decision making tool for the food industry? Int J Food Sci Technol 37:339–349. doi:10.1046/j.1365-2621.2002.00547.x

Enghagen LK (1990) Teaching ethics in Hospitality & Tourism Education. J Hosp Tour Res 14:467–474. doi:10.1177/109634809001400250

EU (2008) Recommendation of the European Parliament and of the Council of 23 April 2008 on the establishment of the European Qualifications Framework for lifelong learning

FAO (2001) Ethical issues in food and agriculture. Food and Agriculture Organization, Rome

Fieser J, Dowden B Internet encyclopedia of philosophy – Ethics. http://www.iep.utm.edu/ethics/. Accessed 8 Mar 2017

Fleddermann CB (2012) Engineering ethics, 4th edn. Prentice Hall, Upper Saddle River

Flynn K, Wahnström E, Popa M et al (2013) Ideal skills for European food scientists and technologists: identifying the most desired knowledge, skills and competencies. Innovative Food Sci Emerg Technol 18:246–255. doi:10.1016/j.ifset.2012.09.004

Goldie J (2000) Review of ethics curricula in undergraduate medical education. Med Educ 34:108–119. doi:10.1046/j.1365-2923.2000.00607.x

IFA (2016) European quality assurance for food studies programmes. ISEKI Food Association. https://www.iseki-food.net/webfm_send/2361. Accessed 18 May 2016

IFT (2016) 2011 resource guide for approval and re-approval of undergraduate food science programs. Institute of Food Technologists. http://www.ift.org/~/media/Knowledge Center/Learn Food Science/Become a Food Scientist/Resources/Guide_Approval_UndergradFoodSci.pdf. Accessed 18 May 2016

Illingworth S (2004) Approaches to ethics in higher education. Learning and teaching in ethics across the curriculum. In: Philosophical and religious studies subject centre. Learning and Teaching Support Network, Leeds

Johnson J (2009) Chapter 11: teaching ethics to scence students: challenges and a strategy. Educ Ethics Life Sci 17:197–213

Kaltoft P, Sandøe P (2001) Teaching ethics in agricultural university. In: Pasquali M (ed) Ikke angivet, pp 1–4

Kohlberg L (1976) Moral stages and moralization: the cognitive-developmental approach. Moral Dev Behav Theory, Res Soc:31–53

Kostaropoulos A (2012) Food engineering within sciences of food. Int J Food Stud 1:109–113

Kroesen O, van der Zwaag S (2009) Teaching ethics to engineering students : from clean concepts to dirty tricks. Philos Eng An Emerg Agenda:227–237

Kuhse H, Singer P (2009) What is bioethics? A historical introduction. In: Kuhse H, Singer P (eds) A companion to bioethics, 2nd edn. Wiley-Blackwell, Chichester, pp 3–11

Layman E (1996) Ethics education: curricular considerations for the allied health disciplines. J Allied Health 25:149–160

Littig B (2004) The Neo-Socratic Dialogue (NSD): a method of teaching the ethics of sustainable development. In: Galea C (ed) Teaching business sustainability: vol. 1, from theory to practice. Greenleaf Publ, Sheffield

Loe TW, Ferrell L (2001) Teaching marketing ethics in the 21 century. Mark Educ Rev 11:1–16

Loui MC, Gunsalus CK (2006) Proposal to the Ethics Education in Science and Engineering Program, National Science Foundation: Role-Play Scenarios for Teaching Responsible Conduct of Research

Lynn C (2010) Teaching ethics with an integrated online curriculum. J Hosp Leis Sport Tour Educ 9:123–129. doi:10.3794/johlste.92.286

Magalhães-Sant'Ana M, Olsson IAS, Sandøe P, Millar K (2010) How ethics is taught by European veterinary faculties: a review of published literature and web resources. In: Romeo Casabona CM, Escajedo San Epifanio L, Emaldi Cirión A (eds) Global food security: ethical and legal challenges. Wageningen Academic Publishers, Wageningen, pp 441–446

Moreno JD (2009) Ethics committees and ethics consultants. In: Kuhse H, Singer P (eds) A companion to bioethics. Wiley-Blackwell, Chichester, pp 573–583

Mossop R (1997) With great difficulty: the assessment of attitudes and ethics at work and in training. Centre for Vocational Assessment Research, Chatswood

Newholm T, Newholm S, Shaw D (2015) A history for consumption ethics. Bus Hist 57:290–310. doi:10.1080/00076791.2014.935343

Noel CZJ, Hathorn LG (2014) Teaching ethics makes a difference. J Acad Bus Ethics 8:1–31

Peer KS, Schlabach GA (2010) Uncovering the moral compass: an integrated ethics education approach transcending the curriculum. Teach Ethics:55–74

Pollan M (2008) In defense of food: an eater's manifesto. The Penguin Press, New York

QAA (2009) Subject benchmark statement – agriculture, horticulture, forestry, food and consumer sciences. Quality Assurance Agency for Higher Education. http://www.qaa.ac.uk/en/Publications/Documents/Subject-benchmark-statement-Agriculture.pdf. Accessed 18 May 2016

RAE (2011) An engineering ethics curriculum map. The Royal Academy of Engineering. http://www.raeng.org.uk/policy/engineering-ethics/ethics. Accessed 8 Mar 2017

Schwartz B (2009) An innovative approach to teaching ethics. J Dent Educ. doi:10.1207/s15327655jchn0504_5

Shani A, Belhassen Y, Soskolne D (2013) Teaching professional ethics in culinary studies. Int J Contemp Hosp Manag 25:447–464. doi:10.1108/09596111311311062

Sims R (2002) Business ethics teaching for effective learning. Teach Bus Ethics 6:393–410. doi:10.1023/A:1021107728568

Thompson KR (2006) A conversation with Steven Kerr: a rational approach to understanding and teaching ethics. J Leadersh Organ Stud 13:61–74. doi:10.1177/10717919070130020501

UN (1948) Universal Declaration of Human Rights

Weaver K (2007) Ethical sensitivity: state of knowledge and needs for future research. Nurs Ethics 14:141–155. doi:10.1177/0969733007073694

Weber J, Glyptis S (2000) Measuring the impact of a business ethics course and community service experience on students' values and opinions. Teach Bus Ethics 4:341–358. doi:10.1023/A:1009862806641

Weber J (2007) Business ethics training: insights from learning theory. J Bus Ethics 70:61–85. doi:10.1007/s10551-006-9083-8

Weston A (1997) A practical companion to ethics, 3rd edn. Oxford University Press, New York

Wharton C (2011) Food beyond nutrition. Teach Ethics 11:15–24. doi:10.5840/tej20111123

Zahn R, de Oliveira-Souza R, Moll J (2011) The neuroanatomical basis of moral cognition and emotion. From DNA to Soc Cogn:123–138. doi:10.1002/9781118101803.ch8

Zwart H (2000) A short history of food ethics. J Agric Environ Ethics 12:113–126. doi:10.1023/A:1009530412679

Chapter 2
With a Little Help from Bioethics

Adelino M. Moreira dos Santos

2.1 A Little Bit of History

2.1.1 Bioethics

Ethics comes from the Greek word "ethos" referring to a particular group of habits distinguishing an individual and, later, to character. It mainly referred to what was considered to be the right way of living, although the meaning of "what is right" or "rightness" was in no way limited to that expressed in a code of values.

Today, it focuses on human action. It encompasses the philosophical study of human behavior, examining individual human and collective behavior in an attempt to distinguish good from evil.

Bioethics is a quite recent discipline. It took shape near the end of the twentieth century mainly in response to advances in biomedicine and because of the more detailed information of the body offered by biotechnology and subject of application by/for health professionals.

It is the discipline that aims to study the morals of an action. Often, it is called moral philosophy. This last statement is incorrect because its main goal is, in fact, to propose the best solution for real dilemmas and not moral distresses.

Alfaro-LeFevre (2011) believes that an ethical-moral dilemma involves a situation where there are various available options, none of which seem satisfactory, although one must decide and choose the best option.

So, we can say that there is an ethical dilemma when healthcare professionals raise the following questions: (a) is an intervention ethically appropriate (b) or what possible interventions should be favored from the ethical point of view to achieve

A.M. Moreira dos Santos (✉)
Coimbra Health School of the Polytechnic Institute,
Rua 5 de Outubro, 3046-854 Coimbra, Portugal
e-mail: adelinosantos@estescoimbra.pt

© Springer International Publishing AG 2018   25
R. Costa, P. Pittia (eds.), *Food Ethics Education*, Integrating Food Science and
Engineering Knowledge Into the Food Chain 13,
https://doi.org/10.1007/978-3-319-64738-8_2

the purposes? The questions behind the decisions are thus constant in the lives of people although the recognition of the undesirable: to decide or not to decide.

The ethical dilemmas that are most often found in health services, based on professional experience and review of the literature, are the continuity of care, meeting a patient's health needs in a humane manner and the allocation of existing resources. Ultimately, the majority of clinical decisions in healthcare are decisions about resources concerning means of consumption and whether they are proportionate or disproportionate.

2.1.2   Decision-Making

Decision-making is a multifaceted cognitive process that is a requisite for problem solving. It is based mainly on a choice among alternatives. Ethical decision-making must reflect different bioethical issues related to the beginning, course, and the end of human life while retaining a multidisciplinary perspective, which considers the philosophical, legal, and scientific aspects.

Any model for ethical decision-making recognizes that it must embrace a systematic set of principles that motivate and guide actions. These principles must be explained and justified, and actions are carried out by the objectives or fundamental rights of others for the care they need.

When considering the advantages of using a decision-making model, among systematized models for ethical decision compiled from the literature, it is necessary to be aware that this represents only an aid and as common sense "a guide to solving ethical problems we face... the same solution is not always the correct one to solve a problem in similar situations."

Some authors (Post 2004) present a reference model at the diverse platforms (at schools, companies, hospitals, civil needs, etc.) of approach to problem solutions. All decision-making models in ethics should consider as many viable options as possible and assess the consequences of each of these before deciding on a course of action. One must be alert to the fact that, as noted above, it is necessary to be aware that doing nothing, to abstain from acting, also indicates a decision has been taken.

Seeking a deeper, though not exhaustive, application that may be used with a minimum of preparation based on critical thinking and reflection for decision-making, we can see that this process in the ethical area follows a flowchart of aid regardless of the approach or model used:

- To identify and formulate the dilemma clearly
- To collect relevant information and consider possible courses of action
- To develop and compare alternative solutions for new problems without any assessment
- To analyze the advantages and weaknesses of each alternative and select an alternative among those that seem best, testing the effects of that choice, i.e.,

evaluating the solutions regarding adaptation to the situation, its viability over time and context
- To plan and implement the solution
- To watch and compare the results to the desired outcome

2.1.3 Relationship with the Service Receiver

The relationship between the professional and the public is unquestionably linked to the need for service and necessarily passes through communication, which can limit its flow.

Ethically, professionals should pay attention to "dangers" to foresee the human needs of the population. This must also be kept during food manufacture until the end when the public is going to form an impression of fabrication procedures. Moreover, we are also talking about respect of identity, dignity, justice, choice autonomy, etc., that requires a considerable amount of thinking and commitment aligned with the Universal Declaration of Human Rights which establishes the right of respect for human dignity, promoting the spirit of brotherhood as a behavior pattern.

Only this process can ensure that these rights are, by principle, systematically respected. In practice, does it mean to respect the people? Well, it is to try to recognize the other as similar: if we cannot accept the fundamental sameness between us, we cannot truly communicate with anyone. This is true and essential whether one may be black and the other white or one starved and the other satisfied.

Of course, this situation is even clearer when the other is someone next to us, for example, a relative. Tomorrow it could be our turn to claim for justice in food distribution. This means that being aware of our own humanity one also recognizes the very real differences between individuals. In a way, it is through my own humanity that I know what each person feels.

To take this suggestion seriously is to turn it into reality. Living in a time of rapid society change, we must avoid a simplistic approach and use caution in practice with our neighbors, for the sake of our responsibility to future generations. These goals are well within our scope since as human beings we naturally feel the duty of responsibility and respect for others and, ultimately, the need for our own sense of rectitude.

Regardless of the remarkable progress, our society still compels us to take decisions through systematic selection and to make judgments. However, a social policy not infrequently causes serious imbalances in the global economy because of limited resources. Concerning the complexity of coexistence of diverse human cultures and interests, the only fair policy follows the footsteps of social and technical evolution. Only a robust and permanent reflection can prevent an increasing amount of indignities and inhumane situations.

2.2 Food Professionals and Bioethics

Food industry products affect everyone, and only a few things are more important to us, especially when dangers are hanging over its proper functioning. This is one subject where our actions and attitudes affect the vital interests of other persons. While we perceive these interests as important in themselves, it means nothing if they are not viewed as essentially ethical or moral concerns. Actual food processing, storage, and distribution should direct our attention to how this concern is taken.

On the contrary, there is a common mindset of concern related to professional performance that is propagated by the media. This is usually done using superficial presentations, but the media provides platforms where everyone feels qualified to take part, give an opinion, and express a point of view on the issues appraised. Furthermore, there is a discourse of review and analysis in the art of ethical and medical issues within the scientific community. By its scientific nature, among high-level nutrition, dietetics, and food engineering education, the latter discourse is inevitably much less accessible than the popular one. Neither obscure nor irrelevant, it is widespread among peers, which incorporate a substantial part of its analysis into scientific production. However, it is reasonable to assume that these two message types, popular and academic, should share the aim of helping to promote understanding of the ethical dimensions to public in general.

2.2.1 Ethical Analysis

Most of the people do not agree on the nature of ethical analysis or the type of expertise required for it. The proper approach is based on the analysis which contains no ethical-moral authority. Both supporters and critics alike doubt the legitimacy of authority and differ on which of the current objection bases can be ignored.

Another important point is that the ethical review should not take sides in any particular case. Instead, it should make clear the moral and conceptual assumptions of a particular point of view. This would make it easier to determine what is a point of disagreement and if a particular moral position is consistent. It may produce a particular moral position more plausible than before the analysis. The choice of topics to consider, the scenarios of the analysis, and the analyst's evaluative paradigm can influence the way the parts of a position are taken into account in a specific matter.

Note that proficiency in the ethical analysis is merely a competence and is not in itself a moral virtue. The analyst is not the only one with moral wisdom. Ethical analysis is a resource that can be used in the analysis of moral reflection on an issue but is not a substitute for reflexive thought on that issue.

In fact, the ethical analysis does not develop suddenly. The main concerns of ethical analysis such as the nature of the action or intention or the importance of duty and the consequences and the virtuous nature of the agent have been debated for many years as part of the philosophical concern with the study of ethics. It is clear that many of these historical developments in ethical thinking are still a focal point and, of course, can also be found in the ethical analysis within the food industry.

All these approaches tend to aim the action topic: what to do in a given situation (main concern of judgment and moral thought). Another perspective is that which concerns the main issues of morality not "what I do" but "how should I do." Advocates of this attitude have carried over the decisive element in a moral problem to some kind of character virtues which are called for in a particular situation.

Apart from the appeal of various religious forms, the following are the main branches of moral thought that, significantly, found expression in the modern literature of ethical analysis, respectively, through the work on utilitarianism (Singer 1993), theories of duty (McCormick 1997), the principlism theories of Beauchamp and Childress (2009), and rights-based approaches of Dworkin (1993).

We can also add a renewed attention to other approaches, some of which are clearly spelled out in the field of ethical analysis: Ethics Narrative (Tovey 1998), Interpretative Ethics (Post 2004), Case by Case (Post 2004), Clinical Ethics (Jonsen et al. 1982), and the Ethics of Virtues (MacIntyre 2007).

Without deep investigation, it should be recognized that the limitless ethical reflection varies in its statement of values and prominent purposes. Moreover, the importance derived from the result of determination is inevitably ethically influenced by all the moral values that can contribute inevitably to action.

Respect for individual autonomy is the dominant value in the ethical approach. Therefore, the choices and decisions must be clearer about public desired objectives.

A difficulty also arises because people do not always have special knowledge about food or even which food is best for them. After all, people compete for limited resources and food workers legitimately exercise their personal autonomy and act according to public health or simply to markets demand.

2.2.2   Bioethics and the Food Industry

Bioethics concerns the welfare and potential suffering of people. These situations appear not to arise from a conflict of ethical choice or values, but only moral values.

We can take ethical situations as a simple, practical exercise, or as marginal ethical problems, including another factor: self-interest as an important point in the competition for society resources.

Food professionals understand and experience ethical problems within their professional community. However, some areas of ethical reflection are naturally more generic. For example, some ethical issues are shared by several differentiated workers such as nutritionists, dieticians, food engineers, chefs, etc. They include problems such as accomplishing with filthy facilities and delivered bad condition food, among several others.

In some cases, these problems do not immediately become a dilemma because they have already fallen under the guidelines both of professional codes as well as the legal framework and so they merely become individual moral stress.

Bioethics relates not only to matters of life and death but also to how each one should live and interact with others in everyday life. The same can be said of professionals concerned with the act within the law when other ethical values are at stake and when they identify ethical problems in their usual work.

In the health area, many upper-level courses (and even some mid-level) include some form of training in bioethics and/or deontology. Moreover, professional structures, both national and international, recognize that curricula should include these concepts so that professionals can have a qualitatively better education covering various areas of knowledge to obtain skills, knowledge, and competence.

Admission to a profession confers the right to practice and the responsibility to ethical practice. If freedom is a measure of independence, it comes with inherent responsibility: you are responsible for decisions that ultimately will support society's well-being.

Some EU ethics projects, for example, integrate Internet tools for vocational orientation, with a particular reference to ethical considerations. In this way, the respective project Ethics Committee recognizes (1) the inherent duality in ethical practice that requires the professional body to protect the general public from the negligence of its members and (2) provides support to its affiliates on ethical issues.

Also valuable is an informal group or even one-to-one discussion, which can be very effective in combating ethical dilemmas and ensure continuous professional development.

2.2.3 Practice of the Profession

This leads us to practice. In an ethical perspective, it should be considered the sociological delineations of the profession area regarding higher education, formal certification, full-time work to earn a living, the organization, and, of course, the Code of Ethics (see Chap. 6).

Using the term "practice," in its primary sense, shows that our concept will not identify jobs with groups of people, but rather with certain professional practices. For example, logically, by carpentry, we would not mean what carpenters do, but the activities of the professional carpentry practice. Therefore, it is expected that the ethics of a carpenter work, the essence of the practice of carpentry, that is, the spirit of the profession, lay down the principles and rules of conduct.

In the same way, it should not be understood, for example, that professionals have no influence as regards the nature of food industry work. In fact, they can change the nature of their profession, as any other professional community can do through their own practice.

On the other hand, professional ethics has to do with the professional practice of compatibility, with the depth of social involvement. Of course, we are particularly interested in the values present in a democratic climate, which reflects the core moral principles of democracy: human dignity, justice, and equity.

The exercise of a profession in the context of civil society in a democracy is limited by some of the fundamental moral principles of that society, accepting, or even improving their practical manifestation, within the personal autonomy.

This concept of personal autonomy rests upon two assumptions: firstly, to have gained autonomy, one must know how to use reason, for example, courses on a plan of action, the choice of activities, the selection of proper establishment of practices, or the continuity of lifestyle. Secondly, a person has to be truly free, not coerced, not dependent upon the intentions of others, but able to make decisions and act which begs the question: how free is a food worker while reasoning as an active participant in the food industry?

Moreover, social autonomy is exercised by a professional community only when it is independent of its design and the limitations imposed by the normative nature of the social context. So, if most of the ethical "duties" can be interpreted and "solved" with the aid of the Code of Ethics, it is not enough, because it means that professionals should not fall back on formulas to address the problems of professional ethics. As individual workers, they have to take this responsibility seriously, maintaining high standards, if they want to ensure public trust.

This is important because management is more likely to intervene and impose the regulation of a profession if there is dissatisfaction or distrust in the efficacy of professional self-regulation.

Essentially, professionalism is the relationship between freedom of professionals to practice and co-responsibility for ethical practice, and all workers have a vital role in all forms of labor education, along with ethical guidance.

2.3   Should We Comply with the Deontological Codes?

2.3.1   Code Structure

Very often, the first application effect of a Deontological Code is to build a professional identity characteristic of each group, thus serving to strengthen professional ties and motivation and boost a feeling of honor of membership in a group.

As for its content, the code should reveal the essential rules to follow, such as the prohibition of harming human dignity, and also the rules that the professional or employer want to respect strictly for business purposes.

Throughout the last century, technical progress has significantly increased the possibilities of intervention on the body. Mainly, since the 1960s this progress has led not only to care but also acts outside of any pathology, such as in ultrasound research for determination/selection of gender in fetuses. Various responses have been suggested to frame the use and development of these technologies.

The abandonment of the moral subject itself is a consequence of the lack of objectification of what is good and what is evil or, even, a concern for dignity, resulting in moral relativism. That is, acting according to our convenience, we may view a subject according to a particular use. However, in health care it is impossible to separate the profession from the vocation, foreseeing good performance in the tasks benefiting the person as a whole, because this relationship reveals what the professional is as a person.

Nowadays, in health technologies, to reach the high professional standards, it requires a functional behavior based on professional competence which is characterized, in addition to implementation capacity, by attitudes that mark the professional as not merely a performer. As professional competence is the result of knowledge integration in a given context, the actual performance is differentiated between the top-level health workers.

A Deontological Code should be the principal instrument of rules appropriate to the function of the work. First, it should be part of the necessary development of professional competence; second, it should establish the designation of the qualities characteristic of professional attitudes. Thus, relatively concise and without entering into the particularity of cases, it could be a professional unifier through its wider recognition.

The term deontology comes from the Greek "deon" meaning among other things, duties. Thus, work ethics would be a system of rules and ethical obligations to which the individual must follow while searching for the correct action that respects the integrity and dignity of every human being.

From the Kant's analysis (Kant 1980) on the act of duty, we understand that the moral nature of an action must reside, not in the ends to which we propose but the guiding principles of the action. Why is this? Because only then may it reach universal objectivity to liberate it from relativism. In fact, to this author, the moral action is that which has universal validity: to discover if my act is, or is not, moral, I should only ask myself if the guiding principle of my action is valid for all humanity.

The regulatory essence that the moral law calls for results in the conflict between what each of us, as an individual, feels and what reason requires us to do. That is the conflict in a person influenced by trends and subjective interests that may deny the status of a person endowed with reason, which is immoral.

As such, deontology stands for professional ethics, implying that members respect rules for action guidelines. Although based on rules for establishing procedures, from the distinction between morals and law, we have the moral judgment of each one to ensure the proper conduct, generally without defined boundaries and contributing thus to the communion of identities, contrary to law, as a social form of rules, which have a circumstantial limit of territory.

As for deontology, the reflection on professional performances should begin before the professional practice. However, when the professional takes office, he is contractually bound to established legal and professional standards. This implies a voluntary agreement to a set of established rules as the most suitable for the exercise of one's profession. Even a precarious labor condition does not negate the responsibilities of belonging to a particular professional class; it means that even new professional dimensions are no excuse for lack of reflection on current praxis.

2.3.2 Some Reasonable Doubts

Am I a good professional? Do I execute my work properly? These are relatively easy questions to answer since they are normal. However, there are other aspects that professionals, although not provided for in the professional code, are shared by all activities, in particular, those resulting from the application of the Ethics Evaluation Criteria: charity, respect for autonomy, justice, etc.

The unseen, solitary tasks that similarly contribute to the advancement of the work should also be reflected: the food engineer helping to unload the food products can do it purely for generosity and cooperation – going beyond the professional duty, toward goodness, shows that a real professional is more concerned with people than only focused on professional tasks.

A word on the Objection of Conscience – that is, the nonfulfillment of obligations or practices of legally established acts because of one's beliefs that prevent the subject from performing them. Here, the noncompliance, although free from penalty, must be assumed individually, privately, and without prejudice to third parties. It means that the breach should only occur for reasons of conscience, that is, the rejection of the standard should be ethically reprehensible to the worker.

Although mainly addressing religion and the army, this is provided for by the democratic national constitutions which usually stipulate the guaranteed right "to conscientious objection, under law" as well as the Universal Declaration of Bioethics and Human Rights and also in the European Convention on Human Rights, which establishes that "everyone has the right to freedom of thought, conscience, and religion." In fact, there are religious, moral, philosophical, or humanistic reasons that can support the objection. However, they must comply with fundamental requirements, among which are:

• It must substantiate a firm purpose, deeply integrated and not a mere opinion.
• This consideration should work for the subject as an internalized obligation.

Note that the framework of medical activities rests largely on self-regulation through ethical principles, i.e., duties and best practices codified by the profession itself. However, even in the particular case of the food industry, the professional is personally responsible for the task assigned by the company having, as a consequence, the need to preserve their professional independence, thus placing the interests of society above personal preferences, or even of employers' influences.

2.4 A Proposal for Food Ethics Education

Food industry professionals have their interaction focused on the community as a whole. Although many of the correct procedures are outlined in good practice manuals, they do not appeal directly to the human faculty of reflecting principles.

Following our reasoning, we have now to illustrate how these bioethics contents can/should be taught to future professionals in this field of dietetics, nutrition, and food engineering, regardless of the tasks assigned to them.

Considering the basic principles necessary for integration of content and further reflection and applying them, in the course we can mention the various types of more common theories at present: ethical virtues, utilitarianism, consequentialism (Post 2004), and contractualism (Scanlon 1998).

As one would assume, it is not easy to suggest syllabi for courses for which we do not know the actual context of their implementation. However, it should be noted that the vast majority of cases do not translate, in fact, into ethical dilemma inasmuch as they consist of moral anxiety solved by the application of ethical, health, and industrial laws, therefore subject to sanctions according to the interpretation of national "moral law."

Unqualified work, inadequate hand hygiene, toilet inadequacy, defrosting and/or food confection in unsuitable temperature conditions, use of nail polish, uncovered hair, kitchens with small work area, use of raw materials of dubious provenance, no use of sanitized uniform, inadequate preparation or space distribution, use of inadequate transport without refrigeration system, adverse operating conditions for transportation of hot food, failure to use a double boiler system, use of cleaning and disinfection products unsuitable for food, poor cleaning and decontamination of premises, food in unsuitable packaging, occurrence of diseases that may affect the safety of products of animal origin, food storage without daylight or appropriate artificial illumination, cleaners and disinfectants stored in areas where food is handled, material cleaning and disinfecting with sufficient frequency to avoid any risk of contamination, use of non-potable water, and many other unfortunate situations arise, basically, from moral relativism and self-interest. These are problems that reveal a clear flaw in the implementation of legislation in this specialized field of the workplace.

To deal with these and the real ethical dilemmas, as previously alluded to in this chapter, we must establish certain learning objectives and skills to be acquired by students. Thus, the student must acquire knowledge to:

- Recognize the main concepts and ethical and moral theories that ground decision-making.
- Relate their praxis with the legal framework and the resulting impact on job performance.
- Deal with populace and material resources with dignity and respect.

The student must acquire skills to:

- Establish an empathetic relationship with people toward the collective good.

- Apply approaches to anti-discrimination and take into consideration the physical, psychological, social, cultural, and religious needs.
- The student must acquire competences for:
- Take responsibility underlying all professional tasks.
- Obtain informed and clarified consent when applied.
- Show good character and professionally integrate sound professional standards in private life.
- Manage ethically reflecting on human and material resources.

As for teaching methodology, it will be mainly expository with a call for analysis and critical reflection on the agreed contents through interaction in the classroom. Subsequently, the evaluation will be done through two methods:

- Continuous assessment: appropriate interventions in classroom analysis introduced in response to common contextual issues. Here, the student's intended involvement is to be alert to current topics, weighing and consolidating critical reflection on actions arising from the decision-making process.

Participation in lectures allows students to learn the concepts congruently with social intervention and professional practice at the same time and also the accountability for their own educative process.

- Summative evaluation: written test (via long answer/essay) on the subjects listed in the programmatic work plus written work. This type of intervention gives the student great freedom to "show what he knows" both in the area of bioethics and deontology field. It allows an efficient measure of creativity, organizational skills, ability to synthesize, and self-evaluation.

This way we can demonstrate the consistency between program contents and the objectives of the course: theoretical knowledge, written and verbally of addressed issues will qualify for permanent exercise of reflection on scientific advances, leading to the so-called "Quality in the Provision of Services."

It is accessed through services humanization, and its natural vehicle is ethical reflection. Therefore, the student will join an interdisciplinary dialogue through reflection tools, such as the interpretation of the law, moral development, and tolerance and respect, outside any hierarchical structure, allowing the free expression of points view, often in conflict. And so we come to the following proposed syllabus for a minimum of 30 hours of teaching:

- The subject of ethics, ethical dilemmas, structure and origin of terms, deontology, and ethics
- Distributive justice in health systems; equality in access to food
- Ethics as a need for conduct guidance; the importance of decision-making
- The Objection of Conscience in work performance
- Analysis of Universal Declaration on Bioethics and Human Rights and other reference documents
- The informed and clarified consent; the invasion of privacy; clinical secret
- Ethical issues in research; the ethics committees

- Population rights and duties
- Adherence to the codes of professional conduct
- The right to health and the distribution of diet resources
- Autonomy and responsibility in work performance

However, of course, this proposal does not finish here. Certainly, the future will bring new challenges, and different social and economic realities will generate many issues of injustice, discrimination, dehumanization, resource depletion, etc. Recognizing the importance of the fundamental questions that reveal how the values and ethical procedures are communicated to society, we can find potential areas of change toward the collective well-being. That is, the discussion of ethics requires a general understanding of the subject, and each one should avoid the temptation to leave the subjective individual moral values to take ownership of the decision-making.

Due to the complexity of the techniques, also the food industry professionals must have a wide and constantly updated training. Whatever the model adopted, it should constitute a fundamental element in their ethical judgment.

2.5   Conclusions

Deontology is synonymous of professional moral, implying voluntary adherence of members to a set of rules of conduct grounded in the philosophy of specific professional values.

As humans, we feel a duty of responsibility and respect for others and, finally, personal rectitude. However, despite extraordinary progress, our society systematically confronts us with decisions and judgments we cannot avoid.

Far from being intrinsically neutral on the ethics, every work holds ethical questions both intentional and belonging to its constitutive structure. In the complexity of the coexistence of diverse cultures and interests, the policy only follows the footsteps of social and technical developments.

Only constant and permanent reflection can prevent lapses into inhuman situations. The Deontological Code must, therefore, contain certain moral principles: respect for human life, the person and his dignity, morality, probity and professional development, and also professional autonomy.

So, should we comply with the Deontological Code?

Yes – if we belong to a professional group with high standards for the practical exercise. These rules have a moral content, showing moral duty embedded in the text of the law. The public believes in this, and they merit it.

References

Alfaro-LeFevre R (2011) Critical thinking and clinical judgment: a practical approach. Saunders, St. Louis

Beauchamp TL, Childress JF (2009) Principles of biomedical ethics, 6th edn. Oxford University Press, New York

Dworkin R (1993) Life's dominion: an argument about abortion and euthanasia. Harper Colling Publishers, London

Jonsen AR, Siegler M, Winslade WJ (1982) Clinical ethics a practical approach to ethical decisions in clinical medicine. MacMillan Publishing Co., New York

Kant E (1980) Oeuvres philosophiques: Des prémiers écrits à la "Critique de la raison pure". I. Gallimard, Paris.

MacIntyre A (2007) After virtue: a study in moral theory, 3rd edn. University of Notre Dame Press, Notre Dame Indiana

McCormick R (1997) A good death – oxymoron? Bioethics Forum 13:5–10

Post SG (2004) Encyclopedia of bioethics. Macmillan Reference USA, New York

Scanlon T (1998) What we owe to each other. Belknap Press of Harvard University Press, Cambridge

Singer P (1993) Practical ethics. Cambridge University Press, Cambridge

Tovey P (1998) Narrative and knowledge development in medical ethics. J Med Ethics 24:176–181

Part II
Food Ethics Issues

Chapter 3
Sustainability and Ethics Along the Food Supply Chain

Harris N. Lazarides and Athanasia M. Goula

3.1 Introduction

During the last 20 years, an alarming trend has been established; the world population is increasing much faster than our capacity to increase food production. The end result is that there is an increasing deficit in food supply, as it is shown in Fig. 3.1. It is predicted that, by 2030, the growth in global population and the impacts of climate change will increase food production needs by 50%, energy demand by 45%, and water demand by 30% (Beddington 2009).

There is a wide range of reasons for this alarming reality. These reasons include:

1. Loss of farmland due to desertification that is driven by climatic changes, soil erosion, and lack of water
2. Decreasing access to farmland due to competing uses, such as infrastructure expansions and other civilian uses, including large-scale recreation projects (i.e., "safari parks" in Tanzania)
3. Competing uses of farmland for production of oil or other nonfood products
4. Decreasing food production capacities of a given farmland due to unsustainable production, handling, and distribution techniques

Although widely recognized as a major driver of desertification, climatic changes cannot be reversed, unless worldwide actions are universally undertaken and effectively applied; nevertheless, such initiatives are rather difficult to implement, as they are expected to stumble on major application obstacles, which include strongly conflicting economic interests. On the other hand, even if the required actions were

H.N. Lazarides (✉) • A.M. Goula
Aristotle University of Thessaloniki, Department of Food Science and Technology,
Thessaloniki, Greece
e-mail: lazaride@agro.auth.gr

© Springer International Publishing AG 2018   41
R. Costa, P. Pittia (eds.), *Food Ethics Education*, Integrating Food Science and
Engineering Knowledge Into the Food Chain 13,
https://doi.org/10.1007/978-3-319-64738-8_3

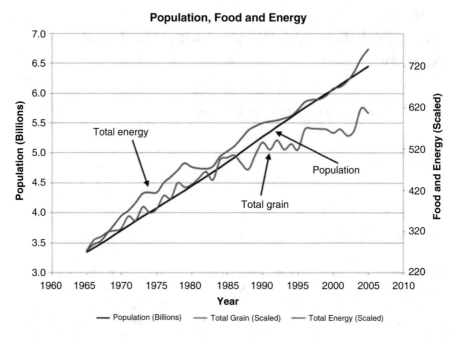

Fig. 3.1 Population and food growth during the period from 1965 to 2005 (Chefurka 2007)

immediately implemented, it would take some decades before the appearance of measurable positive results.

Soil erosion and lack of water are directly linked to climatic changes; yet, they also depend on other factors, such as deforestation, increasing demands for irrigation water, etc.

Decreasing access to farmland is due in large part to competing for non-farming uses, such as infrastructure expansions (i.e., highways, bridges, parks, schools, airports, shopping centers, and other civilian uses). Large areas of farmland may be withdrawn from farming to be used in large-scale recreation projects (i.e., "safari parks" in Tanzania).

During the last decades, the growing use of farmland for the production of nonfood products (i.e., fuel plants) has been adding a substantial pressure on the food production resources. Although fuel plants appear as an attractive alternative as a renewable energy resource, there is clear indication that competing uses of farmland for the production of oil or other nonfood products will further increase shortage of food and its price.

As all other intervention possibilities are rather difficult or nearly impossible to apply, an appropriate focus on sustainability along the food supply chain appears to be a promising field, where our society can expect to draw benefits regarding improved efficiency, increased productivity, environmental protection, and social justice. Lack of sustainability at any stage of the food supply chain is expected to raise severe environmental, ethical, economic, and social issues.

The main objective of this chapter is to present and discuss the importance of sustainability along the entire food supply chain *(from farm to fork)*. As sustainability is firmly associated with certain social concerns and legitimate questions of production and/or consumption ethics, such questions and concerns deserve an appropriate focus.

3.2 Defining Sustainability and Its Association to Ethics

Sustainable supply chain management (SCM) is "the strategic, transparent integration and achievement of an organisation's social, environmental and economic goals in the systematic coordination of key inter organisational business processes for improving the long term economic performance of the individual company and its supply chains" (Carter and Rogers 2008).

Nevertheless, the term "sustainable development" is socially and politically constructed, just as "sustainable agriculture" is a slippery and broad-ranging term. In launching their "Sustainable Farming and Food Strategy" in December 2002, following recommendations contained within the Curry Report, the Department for Environment Food and Rural Affairs (DEFRA) defined "sustainable development" as "a better quality of life for everyone, now and for future generations to come" (Ilbery and Maye 2005). While similar to the original Brundtland Report definition (Brundtland 1987) and simple in principle, it is difficult to achieve in practice because for DEFRA it means meeting four objectives at the same time:

- Social progress that recognizes the needs of everyone
- Effective protection of the environment
- Prudent use of natural resources
- Maintenance of high and stable levels of economic growth and employment

There is no precise definition of what sustainable agriculture will comprise, despite the plethora of studies about the future of the planet about food production. As Francis and Hildebrand (1989) state: "everyone assumes that agriculture must be sustainable. However, we differ in the interpretations of conditions and assumptions under which this can be made to occur." There is no agreement on how to operationalize and measure sustainability (Nousiainen et al. 2009). Indeed, sustainable development is an umbrella term covering many aspects and is filled in by various activities and practices (Van Lente and Van Til 2008). So, there are many interpretations but not a consensual view on what a sustainable agricultural sector should look like (Zwartkruis et al. 2012).

Nowadays, sustainability is often seen as the combination of three factors: economic, social, and environmental. With a triangular diagram, the relationship between sustainability and the three dimensions could be explained in a semiquantitative way (Fig. 3.2). Point 1 in the diagram represents an organization that pursues purely economic goals; Point "a" is one that has purely social objectives and Point "b" pure environmental goals. As time passed, understanding of the importance of

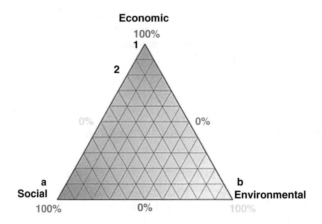

Fig. 3.2 The three dimensions of sustainability explained on a ternary diagram

social issues grew, and companies moved to Point 2. In those times, natural resources were virtually unlimited and the disposal of effluents did not present a problem. However, as industries got surrounded by cities and rivers became more polluted, environmental regulations were enforced, and companies moved to Point 3 in the diagram.

The weakness of portraying sustainability as a mix of three dimensions is that they are all taken as interchangeable and of equal weight. The social and economic systems are subsystems of the environmental system. In this context, a food company takes energy, water, minerals, and land from the environmental system; labor from the social subsystem; and capital from the economic subsystem. In return, the food company produces income for the employees and food for the social subsystem; and the food company and employees pay taxes that benefit the economic system.

The food industry has many impacts on sustainability and vice versa (Maloni and Brown 2006). Discerning customers are increasingly interested in the origin of food products, what they contain, and who made them. Also, policy makers, legislators, influence groups, and financial institutions are progressively placing pressure on firms to report on sustainability performance (Kolk 2004). As Flint and Golicic (2009) observe, not surprisingly, sustainability has received increasing attention in the literature as a potential differentiating competency for supply chains and has become an inescapable priority for firms worldwide (Bourlakis et al. 2014).

3.3 The Food Supply Chain

The food supply chain (FSC) involves a broad range of human activities extending from farm to consumer table. Raw material production, food processing, food storage, food distribution, and consumer handling constitute successive stages along the FSC (Fig. 3.3).

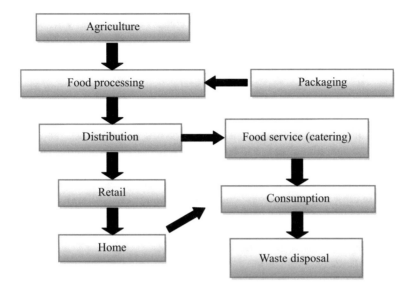

Fig. 3.3 Main stages in the food chain

The food supply chain is complex and includes different kinds of actors who produce, process, or distribute food and agricultural products. Thus, coordinating change is a daunting task. The actors are interdependent, so if something changes in one part of the chain, it will affect the other parts as well. This condition compels innovation to take account of the different interests and ideas; that is, the change should be systemic.

The food industry is a very dynamic industry with constant changes in customer demands (Beske et al. 2014). This calls for the ability to quickly adapt strategies and reconfigure resources, precisely the requirements for which the dynamic capabilities concept has been posited (Zhu et al. 2012). In the modern food industry, processes that have become industrialized are characterized by mass production. Furthermore, production, financing, and marketing have become internationally integrated to form global food supply chains (Trienekens et al. 2012). Such food supply chains can be defined as "a set of interdependent companies that work closely together to manage the flow of goods and services along the value-added chain of agricultural and food products, to realize superior customer value at the lowest possible costs" (Folkerts and Koehorst 1998). Globalization, along with changing marketing techniques, consumption trends, and modern technology, has simultaneously raised concerns in regard to the economy, society, and the environment. Safety and quality are of utmost importance in the food industry, whereas the need for collaboration and coordination is often emphasized (Beske et al. 2014).

Since the 1950s, the food supply chain sector in industrialized countries has grown dramatically. Increasingly, this growth has been challenged by environmental problems, such as emissions of polluting substances to the air, soil, and ground water, and by societal concerns about nature conservation, animal well-being, and quality and safety of food products (Bos et al. 2003; Zwartkruis et al. 2012).

The economic conditions for the food supply chain have changed as well. Due to globalization, for example, consumers can now choose between many food products from all over the world, and they have higher demands regarding quality, traceability, and environmental friendliness of products and processes (Ruben et al. 2006). According to some scholars, it is no longer the producer, but the consumer who determines the rules of production and marketing of food (Bekke and De Vries 2001). As the agricultural sector is facing a lot of technical, economic, sociopolitical, and ecological challenges, innovations are seen as inevitable (Werrij 2007). Given the concerns related to animal welfare, biodiversity, and environmental problems, the interest in sustainable practices is increasing (Pretty 2008). Furthermore, innovations in the food supply chain are needed to maintain the "license to produce" (social problem), the "license to operate" (policy problem), and the "license to deliver" (market problem) (Zwartkruis et al. 2012).

Different research methods have been adopted to help develop different theories of sustainable production and food supply chain management to promote efficiency, value, and quality (Dey et al. 2011; Kuik et al. 2011; Wiese and Toporowsky 2013); in addition, a wide range of cases have been reviewed, to examine sustainable practice (Brammer and Walker 2011; Christopher et al. 2011; Spence and Bourlakis 2009). However, a gap remains between theory and practice about the significance, importance, and application of sustainable practices.

Food production and consumption can have both a positive and negative effect on the environment, and much of the analysis of this is well covered in the environmental and agricultural economics literature; but it can also have an impact on consumer health, social inclusivity, job satisfaction, animal welfare, and a variety of other sustainability indicators. Social and environmental sustainability in food supply chains remains high on the political and economic agenda and reflects the missions of the International Standard Organization (ISO) and World Trade Organization (WTO). However, there is still some way to go throughout the food supply chain to improve its contribution to this wider sustainability agenda, both to regain and retain consumer trust, but also to take advantage of additional opportunities for generating profit (Svensson and Wagner 2012) (Table 3.1).

Unless each and every stage of food supply chain is based on sound sustainability principles, the FSC cannot be characterized as truly sustainable; thus, an effort will be made to discuss all aspects of sustainability for each and every separate step/stage along the FSC.

Table 3.1 Sustainability indicators

Life cycle stage	Indicators		
	Economic	Social	Environmental
Origin of resource – seed production, animal breeding	Degree of control of seed production/ breeding	Diversity in seed collecting options Degree of cross-species manipulation	Ratio of naturally pollinated plants to genetically modified/hybrid plants per acre Reproductive ability of plant or animal % of disease-resistant organisms
Agricultural growing and production	% return on investment Cost of entry to business Farmer savings and insurance plans Flexibility in bank loan requirements to foster environmentally sustainable practices	Diversity and structure of industry, the size of farms, and no. of farms per capita Hours of labor/ yield and/income Avg. farm wages vs. other professions No. of legal laborers on farms, ratio of migrant workers to local laborers, no. of animals/ unit, time animals spend outdoors	Soil microbial activity Quantity of chemical inputs Air pollutants Water withdrawal vs. recharge rates No. of contaminated bodies of surface water % waste utilized as a resource Energy input Ratio of renewable to nonrenewable energy Portion of harvest lost due to pests, diseases
Food processing, packaging, and distribution	Relative profits received by farmer vs. processor vs. retailer	Quality of life and worker satisfaction	Energy requirement Waste produced % of waste and by products utilized in food processing industry % of food lost due to spoilage/ mishandling
Preparation and consumption	Geographic proximity of grower, processor, packager, retailer Portion of consumer disposable income spent on food	Nutritional value of product Food safety Rates of obesity disease/conditions Health costs from diet related Balance of average diet	Energy use in preparation, storage, refrigeration Packaging waste/calories consumed Ratio of local vs. nonlocal and seasonal vs. nonseasonal consumption
End of life	Ratio of food wasted to food consumed	Ratio of food wasted vs. donated to food gatherers	Amount of food waste composted vs. sent to landfill/incinerator/ wastewater treatment

Reconstructed from: Heller and Keoleian (2003)

3.4 Aspects of Sustainability Along the Food Supply Chain

3.4.1 Raw Material Production

Food market globalization has led to a major shift in food production priorities. High safety, satisfactory quality, and acceptable-reasonable cost used to be the ideal ranking of food production priorities, long before food market globalization was established. Market globalization is only achievable if food production costs can be severely compressed. Based on this fact, lowering production costs became an absolute prerequisite and a top priority, while acceptable quality and product safety became secondary, less important priorities.

This new priority ranking could only be achieved with production intensification practices, including GMOs, hormones, antibiotics, etc., that go against nature and cannot be considered sustainable. Besides the lack of sustainability, these intensification practices have led to a series of food crises posing severe food safety threats of a new nature and extent, rather difficult to contain and address. They also raised serious questions regarding food production ethics.

The above discussion has shown that intensification practices tend to violate fundamental laws of nature; therefore, they are neither sustainable nor ethical. Man is supposed to protect the natural environment, working with nature (not against nature) to cover his needs, making sure that the same possibility will be offered to his descendants.

If we recognize that "mother nature" is providing us with everything we need to survive, then working against nature is equivalent to "killing our mother"; this is how it came for ancient Greeks to name "mitroktonos" (mother killer) anyone causing severe damage to a natural ecosystem.

On the other hand, intensified, unsustainable practices lead to mediocre or poor product quality, commodity-type produce that is sold at low prices, making it rather difficult for small farmers to survive. This adds another dimension to lack of sustainability, associated with financial extinction of rural populations.

There is obviously a need to develop integrated and biological production schemes that lead to quality differentiation *(improved quality/safety)* and added-value products *(product de-commoditization)*. This is the only way for a small to medium farmer to survive. Small farmers need to cooperate for cheaper supplies, technical support, and direct sales for better prices by "shortening" the supply chain *(minimize middleman intervention)*.

Even if one accepts the theory of quality/safety equivalence of GM food to their conventional counterparts, genetic engineering is posing a series of crucial issues, such as:

1. Impact on biodiversity
2. Impact on food security and food sovereignty
3. Violation of consumer rights (i.e., the right to free, informed choice, based on clear product labeling; the right and duty to protect nature)

4. Impact on the institution of agriculture
5. Bioethics (patenting natural genetic resources)
6. Social control of scientific research

The following fundamental questions can be raised:

1. What is the cost of promised affluence?
2. How far can the humanity reach work against nature?
3. Can we afford non-sustainable production practices in the long run?

How can we secure safe, high-quality food? Based on the above analysis, it becomes evident that food producers need to change production priorities; they need to abandon the problematic scheme of *low cost – acceptable quality – questionable safety* and adapt the desirable scheme of *high safety – satisfactory quality – acceptable cost*. The low-cost scheme serves the short-term interests of the large producers, but it goes both against nature and against consumer interests. The high-safety scheme serves the long-term interests of the producers, while it is in line with environmental protection and serves consumer interests.

Crops may be lost in the preharvest, harvest, and postharvest periods. Preharvest losses occur before harvesting and may be due to adverse weather, insects, microbes, weeds, and other catastrophic factors. Harvest losses occur between the beginning and completion of harvesting, while postharvest losses occur between harvest and consumption.

Depending on available harvesting, transportation, and storage capacity, a large percentage of raised crops (often more than 50%) never reach the consumer plates in developing countries.

3.4.2 Sustainable Food Processing

Food processing uses a limited number of inexpensive, unelastically priced raw materials to produce a large variety of elastically priced end products. Creating food from raw materials adds value to agricultural products; it also develops new and innovative uses that better serve the needs of the consumer. The end result is positive to both ends of the food chain, the primary producer on one end and the consumer on the other.

Sustainable food processing calls on minimizing the use of process utilities (i.e., water, energy, etc.), while maximizing product yields. In both cases, the result is a decreased cost of production together with a positive environmental impact.

Packaging is an essential element for protection and preservation of quality of food products. Besides, it has secondary roles including containment, communication, convenience, marketing, security, and portion control. From the environmental point of view, the ideal packaging would be one capable of being efficiently recycled an infinite number of times or returned to the ground as nutrients after composting. The "recycling loop" in Fig. 3.4 represents the circulation of renewable or

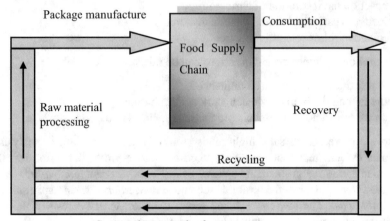

Fig. 3.4 Cradle-to-cradle cycle for food packaging

nonrenewable materials in a closed loop with zero waste. Materials capable of infinite recycling are reused over and over again without undergoing degradation. The "composting" loop cycles biodegradable materials made from renewable resources made from biomass.

With the exception of aluminum and steel, all the other packaging materials are limited to a certain number of recycling cycles, and in most cases, they are downgraded during recycling, meaning they cannot be used in the same type of packaging. However, recycled plastics have broad nonfood applications in automotive, household, and apparel products. Until new materials, which could be infinitely recycled or returned to the ground, are created, saving resources by reducing packaging use, reusing as much as possible, and improving recycling are the only short-term alternatives (Morawicki 2012).

Proper, well-presented food labeling should be informative and offer guidance to well-informed consumer choices, based on best consumer protection and environmental protection.

3.4.3 Sustainable Food Handling/Distribution

Regarding emissions of greenhouse gases, food transportation is on average 11% responsible for the total emissions generated during food production, processing, and distribution in the United States (Weber and Matthews 2008). Emissions from transportation vary by the type of food product (i.e., shelf stable, fresh refrigerated, or frozen), a method of transportation (e.g., sea freight, river vessel, train, truck, or plane), the size of the transporting vehicle (e.g., pickup truck vs. semitrailer truck), and distance. When looking at energy consumption, food transportation accounts for 14% of the total energy consumed during the whole cycle from agricultural production to restaurants and home consumption (Morawicki 2012). Considering the lack of an alternative to fossil fuels in the near future, the only action to minimize the impact of transportation is to become more efficient in energy utilization by technical, operational, and logistical improvements.

Emerging tools for sustainable distribution can offer valuable assistance in minimizing food losses and handling costs. On the other hand, globalization and strong competition for food market control raise serious questions about sustainability and food security across the world. Mishandling of food raises additional questions and jeopardizes food safety.

3.4.4 Sustainable Food Consumption

Consumer responsibility to sustain food quality and safety points to the need for consumer training, as mishandling of food by the consumer will easily cancel every effort and achievement of the food industry with respect to quality, nutritional value, and safety, thus canceling the positive impact of dearly paid achievements in food research, that is, canceling the positive impact of scientific progress.

Questionable consumption ethics and wrong use of food lead to food waste on one hand and health problems (i.e., obesity and associated diseases) on the other. Questionable consumption ethics refer to the lack of fellowship to the needy *(hungry)*; environmental disrespect *(use of overpackaged food items)*; and lack of support for local, traditional food and biological or integrated farming products; i.e., inconsiderate acceptance of GMO and intensive farming products.

Wrong use of food is a serious problem with severe health implications. The so-called "Western" style of eating results in a high intake of calories, sugar, salt, animal fat, trans-fatty acids, and cholesterol, thus leading to "Western" world diseases, such as obesity, diabetes, hypertension, cardiovascular diseases, anemia, certain types of cancer, etc.

Obesity statistics (www.worldobesity.org) point to a major epidemic phenomenon with great socioeconomic impact. Quite often, obesity-triggered diseases lead to death, after a period of medical care that adds a major financial burden on the Medicare system. The question is who should pay for this burden? Naturally, the answer should be those who create it. It is probably time to consider an "unhealthy food" taxation, similar to that applied to tobacco products. Besides the financial support of the Medicare system, such a taxation would probably help the consumers reconsider and – hopefully – correct their eating habits.

There are a number of interventions to be made in a concerted action mode, to reach a positive result toward modifying consumer behavior. Such interventions include:

1. Improvement of food and nutrition education for all (independent of age and education level), including medical students

2. Improvement of public awareness (radio, TV, mass communication media, Internet)
3. Reforming medical education programs to reinforce preventive medical care (prevention vs. curing)
4. Carrying out clinical nutrition studies
5. Establishing corrective nutrition policies at local and state level
6. Enforcing understandable nutritional labeling
7. Legislating and enforcing corporate responsibility

3.4.5 Sustainable Water Use

The overexploitation of water resources, pollution in water sources, changes in regional weather patterns, and an increasing demand for water as a result of population growth are creating the right conditions for a water crisis shortly.

In the food industry, water use can be divided into four groups:

(a) Process water as a primary ingredient
(b) Cleaning water for washing and rinsing fruits, vegetables, and carcasses and for washing crates, bottles, and containers
(c) General-use water to convey product through the process, for product cooling, for cooking and blanching, for product pasteurization, and to sanitize equipment and processing rooms
(d) Utility water for cooling towers, boilers, and scrubbers; ice making; in water lubricated bearings; for general use in processing and administrative buildings; and landscape watering

Figure 3.5 shows the water inputs at each step of the typical food supply chain.

Water used in processing plants produces impacts at local, regional, and global levels due to its removal from local resources (e.g., aquifers); the potabilization process, which requires energy and chemicals; and the generation of effluents. In most cases, effluents of processing plants have high solid contents and need treatment before disposal. Wastewater treatment uses energy and chemicals, thus causing a direct, negative environmental impact due to direct release of greenhouse gases (GHGs), such as carbon dioxide and methane.

So, improving the efficiency of water consumption not only contributes to the preservation of this valuable resource, but also saves money and reduces direct and indirect carbon emissions (NCDENR 2009).

Water reduction programs are unique to each industry and location. However, there are five general strategies that can be followed to minimize the impact of water consumption in most food processing plants (Russell 2005):

• Minimize consumption by adjusting processes and equipment.
• Replace processes with alternatives that consume less water or no water.
• Reuse water in the same process.

Fig. 3.5 Real water input and virtual water flow in the food production chain

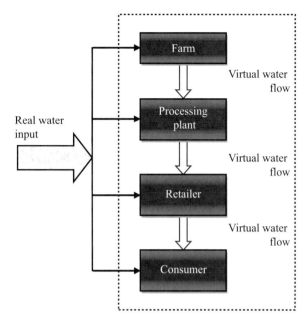

- Recycle water and use in other processes.
- Replenish water by implementing mitigation projects.

3.4.6 Sustainability and Food Waste

Food production and consumption generate food and nonfood wastes. Food wastes are plant or animal materials that were intended for human consumption but ended up being discarded for different reasons. Nonfood wastes include plant and nonedible animal parts, dead animals, manure, and packaging material.

Food waste is a major ethical issue with large dimensions and devastating social impact. About a third of the food for human consumption is wasted globally – around 1.3 billion tons per year (FAO 2009). According to the data in the Eurostat database of the European Commission (EC), the waste generated in 2002 from the manufacture of food products, beverages, and tobacco products was on average about 154 kg per inhabitant and year (Fig. 3.6). According to European Union statistics (http://ec.europa.eu/food/food/sustainability 2013), about 90 million tons of food is wasted annually in Europe – agricultural food waste and fish discards not included.

Food waste in industrialized countries is as high as in developing countries. In developing countries, over 40% of food losses happen after harvest and during processing, while in industrialized countries, over 40% of food losses occur at retail and consumer level. The fact is that for various reasons, food is wasted throughout the entire food chain – from farmer to consumer.

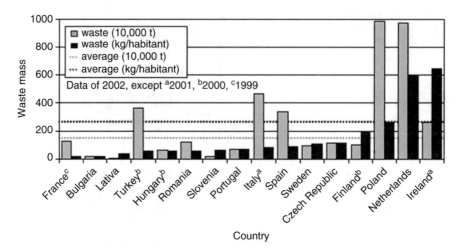

Fig. 3.6 Waste generated in 2002 from the manufacture of food products, beverages, and tobacco products (Eurostat 2005)

Throughout the food supply chain, the minimization of solid waste translates into a direct and indirect reduction in energy consumption, water use, land use, and use of natural resources. The well-known adage "reduce, reuse, and recycle," also known as 3Rs, becomes useful when dealing with both organic and inorganic food and nonfood waste. *Reduce* is about waste generation avoidance. The smaller the amount of produced waste, the lesser handling and treatment it later requires. *Reuse* refers to the reutilization of materials for the same purpose before considering recycling or disposing. *Recycling* is transforming the waste into another usable material (Morawicki 2012).

The EPA (1997) recommends the application of the Food Recovery Hierarchy as a way of reducing food waste going to landfills or incinerators at all stages of the food supply chain. From most desirable to least desirable, the hierarchy uses the following steps:

- Source reduction: reduce the amount of waste generated at any step of the chain.
- Feed hungry people: make available perishable and nonperishable unspoiled food to food banks and shelters.
- Feed animals: supply food scraps to farmers and zoos.
- Industrial uses: send meat and fat trimmings to rendering plants so fat and protein can be separated and used as ingredients for animal feed. Fat can also be used as fuel or as a stock material for the production of soap and other products. Some waste from fruits and vegetables can be used to extract compounds of interest (e.g., nutraceuticals and color compounds).
- Composting and biogas production: use food wastes to make compost or biogas.
- Landfilling/incineration: the last resort to be considered when any of the previous options are not viable.

3.4.7 Sustainability and Energy

Energy is an important input in growing, processing, packaging, distributing, storing, preparing, and disposing of food. It is estimated that the energy consumption of the food chain is responsible for approximately 18% of total primary energy use (Tassou et al. 2014). Domestic operations, cooking, and refrigeration make the highest contribution, followed by manufacturing, commercial transport, agriculture, retail, and catering.

In agriculture, even though the energy consumption is not very high compared to the whole food chain, energy savings of up to 20% can be achieved through renewable energy generation and the use of more efficient technologies and smart control systems. Energy can be saved at the processing plant level by optimizing and integrating processes and systems to reduce energy intensity. Approaches that can be considered include (Campden 2011):

- Design, optimization, and validation of new and modified processes including identification and rigorous assessment and development of more efficient and effective technological approaches for food and drink processing and packing operations
- A better understanding of how processes work and use of knowledge for better process control, better application of automation, and improved process efficiency and flexibility through better scheduling and logistics advanced sensors and equipment for online measurement and intelligent adaptive control of key parameters
- Reduction of processing requirements to improve quality without compromising safety
- Minimization of waste through energy recovery and better use of by-products

In the food retail sector, significant progress in energy efficiency has been made in recent years, but potential still exists for improvements in the efficiency of refrigeration systems, refrigeration and heating, ventilation and air conditioning (HVAC) system integration, heat recovery and amplification using heat pumps, demand side management, system diagnostics, and local combined heat and power (CHP) generation and trigeneration. Energy-saving opportunities also exist for the use of low-energy lighting systems, improvements in the building fabric, integration of renewable energy sources, and thermal energy storage. At home, energy savings can be achieved through the use of more efficient appliances and food preparation methods such as microwave rather than oven cooking, the use of more temperature-stable foods, and changes in consumer diets and behavior.

3.5 Evaluating the Sustainability of Food Supply Chain

An important step toward sustainable business practices is the design and development of a system-based measuring tool providing information on the sustainability of all companies that contribute to an end product. There is a need to assess sustainability in a holistic approach incorporating economic, social, and environmental dimensions (Binder et al. 2012; Singh et al. 2012). For this purpose, indicators can be used that are defined as quantitative measures against which some aspects of the expected performance of a policy or a management strategy can be assessed (Glenn and Pannell 1998). In the last 2 decades, much attention has been paid to establishing indicator lists (Bockstaller et al. 2009; Rinne et al. 2013). However, the selection of these indicator lists is not always clearly described, the lists contain both qualitative and quantitative indicators (Van Passel and Meul 2012), and they do not equally address all three dimensions (Binder et al. 2012; Singh et al. 2012).

Life cycle assessment (LCA) is an analytical method used to evaluate the resource consumption and environmental burdens associated with a product, process, or activity. While the standard LCA method has been applied mainly to manufactured products, the numbers of LCAs evaluating agricultural and food production processes are increasing (Andersson and Ohlsson 1999; Sheehan et al. 1998; Wang et al. 1997). Heller and Keoleian (2003) presented a broad set of indicators covering the life cycle stages of the food system. Indicators address economic, social, and environmental aspects of each life cycle stage: the origin of the resource; agricultural growing and production; food processing, packaging, and distribution; preparation and consumption; and end of life.

Van Asselt et al. (2014) developed a protocol for sustainability assessment of agri-food production systems based on consultations with both a group of scientific experts and a policy maker (Fig. 3.7). The first step of the protocol consists of defining the case study in close collaboration with the policy maker. The second step is to establish a gross list of indicators, categorized in themes within the social, environmental, and economic dimensions of sustainability. These indicators should be measurable, sensitive to variations, relevant to the case study, and related directly to the theme (Bélanger et al. 2012; Dantsis et al. 2010). Consequently, the gross list of indicators is downsized to a list of core indicators that are most relevant for assessing sustainability. In step 4, the list of core indicators is discussed with the policy maker to determine whether it contains all relevant indicators, whereas the next step is the determination of sustainability limits. Limits were derived based on legal norms, policy targets mentioned by the government, or the best performing system. Consequently, values are assigned to each of the core indicators for the policy scenario in the case study based on literature, expert knowledge, and additional calculations. The obtained values for the core indicators as well as their sustainability limits are inserted into a weighing tool to perform an integral assessment of the sustainability of proposed scenarios by normalizing the different indicators. The overall sustainability of a system is then weighed by comparing the three dimensions using importance weights. Finally, the weighing tool is used to visualize the

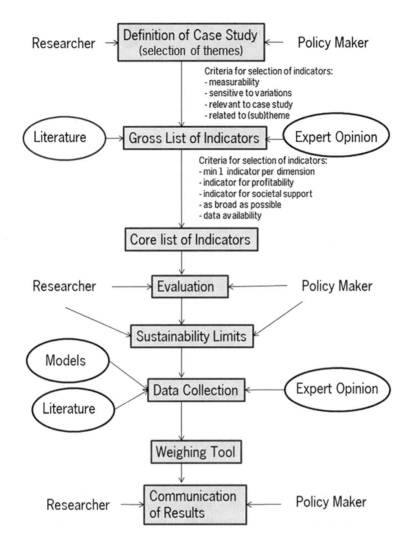

Fig. 3.7 Protocol for evaluating the sustainability of agri-food production systems (van Asselt et al. 2014)

effect of different weights and compensability factors at individual indicator level and the overall sustainability level.

Bourlakis et al. (2014) adapted a performance measurement framework widely used in the supply chain literature (Aramyan et al. 2006; Shepherd and Gunter 2006). This framework allows chain-wide measurement and accommodates the inclusion of nonfinancial measures, which are important in the sustainability context (Shepherd and Gunter 2006). The framework comprises five categories: consumption, flexibility, responsiveness, product quality, and total supply chain. Within this framework, a total of 18 sustainability measures relevant to food industry supply chains were identified from the literature contributions:

- *Consumption*: production/operational/raw materials cost, storage cost, delivery and distribution cost, waste, financial cost, gross profit margin
- *Flexibility*: flexibility in extra volume orders, flexibility in delivering in an extra point of sales
- *Responsiveness*: responsiveness in the arranged lead time, responsiveness in delivery regarding arranged point of sale, responsiveness in delivery regarding the ordered type of product
- *Quality*: quality of the firm's product, product conservation time, consistency of traceability system, storage and delivery conditions, quality of packaging
- *Total supply chain*: firm's perception of its own supply chain performance, firm's perceptions of market opinion regarding its supply chain performance

3.6   Conclusions

Lack of sustainability along the food supply chain is a major cause of questionable production and consumption ethics with severe socioeconomic impact.

Lack of sustainability is raising a range of ethical (moral) issues that challenge our "democracy" and question our respect to human rights.

The world population is increasingly suffering from food in one way or another (hunger, malnutrition, obesity, diet-related diseases).

Globalization and GMOs pose a severe threat to sustainability across the food supply chain, with major negative implications for food security and food sovereignty.

Citizen-consumer education is urgently needed to improve food handling, promote sustainable use of food (minimizing waste), and improve eating habits to fight obesity/diet-related diseases.

Agronomists, food scientists/engineers, dieticians, food policy makers, and communicators are among specialists that need to cooperate in concerted actions leading to sustainable production, handling, and consumption of food.

We, food scientists/engineers, are naturally questioned about our social impact: If we claim to be effective and efficient in improving our food supply system and man welfare, why more and more people do suffer from food?

References

Andersson K, Ohlsson T (1999) Life cycle assessment of bread produced on different scales. Int J Life Cycle Ass 4(1):25–40

Aramyan LH, Ondersteijn C, van Kooten O, Oude Lansink AG (2006) Performance indicators in agri-food production chains. In: Ondersteijn CJ, Wijnands JH, Huirne RB, van Kooten O (eds) Quantifying the agri-food supply chain. Springer, Dordrecht, pp 47–64

Beddington J (2009) Food, energy, water and climate change – a perfect storm of global events? http://webarchivenationalarchives.gov.uk/+/http://www.bis.gov.uk/goscience

Bekke H, De Vries J (2001) De ontpoldering van de Nederlandse landbouw, Het Ministerie van Landbouw, Natuurbeheer en Visserij 1994–2000. Garant-Uitgevers n.v., Leuven-Apeldoorn, Belgium-the Netherlands, p 202

Bélanger V, Vanasse A, Parent D, Allard G, Pellerin D (2012) Development of agri-environmental indicators to assess dairy farm sustainability in Quebec EasternCanada. Ecol Indic 23:421–430

Beske P, Land A, Seuring S (2014) Sustainable supply chain management practices and dynamic capabilities in the food industry: a critical analysis of the literature. Int J Prod Econ 152:131–143

Binder CR, Schmid A, Steinberger JK (2012) Sustainability solution space of the Swiss milk value added chain. Ecol Econ 83:210–220

Bockstaller C, Guichard L, Keichinger O, Girardin P, Galan M-B, Gaillard G (2009) Comparison of methods to assess the sustainability of agricultural sys-tems. A Rev Agron Sustain Dev 29(1):223–235

Bos B, Groot Koerkamp PWG, Groenestein K (2003) A novel design approach for livestock housing based on recursive control–with examples to reduce environmental pollution. Livest Prod Sci 84(2):157–170

Bourlakis M, Maglaras G, Aktas E, Gallear D, Fotopoulos C (2014) Firm size and sustainable performance in food supply chains: insights from Greek SMEs. Int J Prod Econ 152:112–130

Brammer S, Walker H (2011) Sustainable procurement in the public sector: an international comparative study. Int J Oper Prod Manag 31(4):452–476

Brundtland G (1987) Our common future: the world commission on environment and development. Oxford University Press, Oxford

Campden BRI (2011) Scientific and technical needs of the food and drink industry. Campden BRI, Chipping Campden, Gloucestershire. See http://www.campden.co.uk

Carter CR, Rogers DS (2008) A framework of sustainable supply chain management: moving toward new theory. Int J Phys Distr Log Manag 38(5):360–387

Chefurka P (2007) What drives population – food or energy? www.paulchefurka.ca

Christopher M, Khan O, Yun O (2011) Approaches to managing global sourcing risk. Int J Supply Chain Manag 16(2):67–81

Dantsis T, Douma C, Giourga C, Loumou A, Polychronaki EA (2010) A methodological approach to assess and compare the sustainability level of agricultural plant production systems. Ecol Indic 10(2):256–263

Dey A, La Guardia P, Srinivasan M (2011) Building sustainability in logistics operations: a research agenda. Man Res Rev 34(11):1237–1259

Environmental Protection Agency (EPA) (1997) Terms of environment: glossary, abbreviations and acronyms. EPA 175-B-97-001. Washington, D.C. EPA

Eurostat (2005) WAQ1, Generation of waste by economic sector and households, Statistical Office of the European Communities

FAO (2009) www.fao.org/news/story/en/item/20568/icode/

Flint DJ, Golicic SL (2009) Searching for competitive advantage through sustainability. Int J Phys Distrib Logist Manag 39(10):841–860

Folkerts H, Koehorst H (1998) Challenges in international food supply chains: vertical coordination in the European agribusiness and food industries. Brit Food J 100(8):385–388

Francis CA, Hildebrand PF (1989) Farming systems research/extension and the concepts of sustainability. Agron Fac Publ 558:1–10

Glenn NA, Pannell DJ (1998) The economics and application of sustainability indicators in agriculture. In: 42nd annual conference of the Australian agricultural and resource economics society. University of New England, Armidale, pp 19–21

Heller MC, Keoleian GA (2003) Assessing the sustainability of the US food system: a life cycle perspective. Agr Syst 76:1007–1041

Ilbery B, Maye D (2005) Food supply chains and sustainability: evidence from specialist food producers in the Scottish/English borders. Land Use Policy 22:331–344

Kolk A (2004) A decade of sustainability reporting: developments and significance. Int J Env Sust Dev 3(1):51–64

Kuik SS, Nagalingam SV, Amer Y (2011) Sustainable supply chain for collaborative manufacturing. J Man Technol Manag 22(8):984–1001

Maloni MJ, Brown ME (2006) Corporate social responsibility in the supply chain: an application in the food industry. J Bus Ethics 68(1):35–52

Morawicki RP (2012) Handbook of sustainability for the food sciences, 1st edn. John Wiley & Sons, West Sussex, UK

North Carolina Department of Environment and Natural Resources (NCDENR) (2009) Water efficiency manual for commercial, industrial, and institutional facilities, North Carolina Division of Pollution Prevention and Environmental Assistance, Division of Water Resources

Nousiainen M, Pylkkänen P, Saunders F, Seppänen L, Vesala KM (2009) Are alternative food systems socially sustainable? A case study from Finland. J Sustainable Agric 33(5):566–594

Pretty J (2008) Agricultural sustainability: concepts, principles and evidence. Phil Transac Royal Soc B Biol Sci 363:447–465

Rinne J, Lyytimäki J, Kautto P (2013) From sustainability to well-being: lessons learned from the use of sustainable development indicators at national and EU level. Ecol Indic 35(0):35–42

Ruben R, Slingerland M, Nijhoff H (2006) Agro-food chains and networks for development. Issues, approaches and strategies. In: Ruben R, Slingerland M, Nijhoff H (eds) Agro-food chains and networks for development. 6–7 September 2004, Wageningen, pp 1–25, Springer, Netherlands.

Russell SM (2005) New hurdles for water reuse. Watt Poultry, USA, pp 24–27

Sheehan J, Camobreco V, Duffield J, Graboski M, Shapouri H (1998) Life cycle inventory of biodiesel and petroleum diesel for use in an Urban Bus, National Renewable Energy Laboratory, Golden, Colorado. NREL/SR-580–24089

Shepherd C, Gunter H (2006) Measuring supply chain performance: current research and future directions. Int J Prod Perform Man 55(3&4):242–258

Singh RK, Murty HR, Gupta SK, Dikshit AK (2012) An overview of sustainability assessment methodologies. Ecol Indic 15(1):281–299

Spence L, Bourlakis M (2009) The evolution from corporate social responsibility to supply chain responsibility: the case of Waitrose. Supply Chain Man Int J 14(4):291–302

Svensson G, Wagner B (2012) Implementation of sustainable business cycle: the case of a Swedish dairy producer. Supply Chain Man Int J 17(1):93–97

Tassou SA, Kolokotroni M, Gowreesunker B, Stojceska V, Azapagic A, Fryer P, Bakalis S (2014) Energy demand and reduction opportunities in the UK food chain. Proc Inst Civ Eng Energy 167(Issue EN3):162–170

Trienekens JH, Wognum PM, Beulens AJM, van der Vorst JGAJ (2012) Transparency in complex dynamic food supply chains. Adv Eng Informatics 26:55–65

Van Asselt ED, Van Bussela LGJ, Van der Voetb H, Van der Heijdenb GWAM, Trompc SO, Rijgersbergc H, Van Evertb F, Van Wagenbergd CPA, Van der Fels-Klerx HJ (2014) A protocol for evaluating the sustainability of agri-food production systems—a case study on potato production in peri-urban agriculture in The Netherlands. Ecol Indic 43:315–321

Van Lente H, van Til JI (2008) Articulation of sustainability in the emerging field of nanocoatings. J Clean Prod 16(8-9):967–976

Van Passel S, Meul M (2012) Multilevel and multi-user sustainability assessment of farming systems. Env Impact Ass Rev 32(1):170–180

Wang M, Saricks C, Wu M (1997) Fuel-cycle fossil energy use and greenhouse gas emissions of fuel ethanol produced from US Midwest Corn, Center for Transportation Research, Argonne National Laboratory

Weber CL, Matthews HS (2008) Food-miles and the relative climate impacts of food choices in the United States. Environ Sci Technol 42:3508–3513

Werrij F (2007) The changing role of agriculture in Europe and how it affects poultry education and technology transfer. World's Poultry Sci J 63(2):205–211

Wiese A, Toporowsky W (2013) CSR failures in food supply chains – an agency perspective. British Food J 115(1):92–107

Zhu Q, Cordeiro J, Sarkis J (2012) Institutional pressures, dynamic capabilities and environmental management systems: investigating the ISO 9000 – environmental management system imple-mentation linkage. J Environ Manage 114:1–11

Zwartkruis JV, Moors EHM, Farla JCM, van Lente H (2012) Agri-food in search of sustainability: cognitive, interactional and material framing. J Chain Network Sci 12(2):99–110

Chapter 4
The Ethics of Consumption

Anna McElhatton

4.1 The Consumer Society

Tell me what you eat and I'll tell you who you are (Brillat-Savarin 2012).

Throughout history food has always played an important role in human existence and that foods have an effect on health is certainly not a new concept. Hippocrates, the father of modern medicine, some 2500 years ago is said to have adopted the view of "Let food be thy medicine and medicine be thy food", a philosophy that fell out of favour in the nineteenth century with the arrival of therapy using pharmaceutical preparations. However, barely a hundred years later, the idea of the role of diet and, therefore, food components and composition began to resurface yet again.

Food has been described as an "intimate commodity" (Winson 1994). People buy food; they ingest it several times a day. These practices are a meaningful and are a sustained arena of action and interaction that has significant societal effects. Food studies is a relatively recent field of investigation and research that calls on people with backgrounds as varied as urban planning, literature, sociology, bioethics, cultural studies, feminist history and global geography to discover and create points of connection in food (Guptill et al. 2013).

Food is a major part of social life, both experientially and structurally with consumers in developed countries believing that they have every right to consume what they like, when they like, and that their choice to do so is a private matter. Yet there is ample evidence that shows that the health and environmental consequences of how rich developed country consumers eat today seriously impinge on the common good.

A. McElhatton (✉)
University of Malta, Faculty of Health Science,
Block A, Level 1, Mater Dei Hospital, Msida, Malta
e-mail: anna.mcelhatton@um.edu.mt

© Springer International Publishing AG 2018
R. Costa, P. Pittia (eds.), *Food Ethics Education*, Integrating Food Science and
Engineering Knowledge Into the Food Chain 13,
https://doi.org/10.1007/978-3-319-64738-8_4

Food is complex and people frequently and constantly make choices about what, when, and how they eat. Such decisions are made within social and cultural contexts that both enable and constrain those decisions. Additional dimensions of complexity emerge with the question of why so many find it so difficult to eat a "balanced diet". There is no single or predominant "theory of food", but there are clusters of perceptions and ideas that help connect different standpoints on what foods are culturally socially acceptable in varied societies.

In the western world, where "healthy" food is largely constructed within a medical paradigm, "healthy" usually means the well-being of humans, and "healthy food" is frequently narrowly defined in terms of the essential material nutrients for human health. Recent advances in food science and technology, together with increased consumer interest in healthy foods and beverages, have facilitated innovation and promoted a market for food products that claim physiological benefits beyond the need for basic nutrition (Frye and Bruber 2012). Consequently, ethical responsibility becomes all the more necessary as industrialisation of agriculture and food production has caused a profound change to food and the relationship that various stakeholders and society at large have with food.

In the last 10 or 20 years, the consumer has become the focus of extensive debates in many human sciences, including economics, sociology, psychology and cultural studies, to name but a few. There are those who maintain that consumers dictate production and fuel innovation that, in turn, creates new service sectors in advanced economies. This can in some scenarios motivate trends in modern politics (Guptill et al. 2013). Consumers embody a simple modern logic, the right to choose. Yet there are those who contend that the consumer is a weak and malleable creature, easily manipulated, dependent, passive and foolish. Immersed in illusions, addicted to joyless pursuits and in spite of ever-increasing living standards, the consumer, far from being a god, is actually a pawn, in games played in invisible boardrooms. Choice, modelled on the affluent consumer experience, has become the central tenet of many political and ethical discourses. At the same time, there is an increasing awareness among some academics of the ecological limits to the consumerist trend, which are already alarming observers of climate change, raw materials and natural resources such as soil, water and air (Gabriel and Lang 2008).

Mass consumption is now widely seen as having become fragmented into a vast array of highly individualised niche products. This phenomenon of mass production is not limited to affluent communities; indeed it has spread to countries with lower wages and looser environmental and social controls, thus developing novel variants of consumerism in communities with low financial potential. It should be noted that every form of production involves the consumption of resources, and every type of consumption results in some production, even if only waste. Yet, consumption is also *work* – it requires the consumer to put in effort into actual physical or virtual searches through high streets, shopping malls or Internet sites; it involves detailed comparisons and hard choices that demands continuous updating and vigilance with each new acquisition of a commodity or service.

Inequalities among consumers are highly evident, leaving substantial numbers with restricted opportunities and choice to purchase. This is contributing to the fragmentation of consumers' experiences with inequalities occurring not just between societies but actually within the said societies. Given such social gaps, it is difficult to talk about *all* consumption and *all* consumers having the same attitudes or limitations, i.e. as being uniform entities or acting as a unified force, with respect to trends in consumption.

Mass consumption is an intrinsic characteristic of market economy. As individual acts, consumption decisions should be the result of our right to exert our right to free will. In addition, it would be desirable that they were carried out in an ethical and responsible way, balancing not only the objective capacity to satisfy consumer's needs but also the consequences that arise from those choices such as the effects on health, environment and labour work conditions of people involved in the production and distribution process. Information is a basic requirement to be able to choose. But, in spite of the fact that information is acknowledged as a consumer's basic right, its practice effectiveness is not enough yet. Advertising contributes to a large extent to that since it is a communication mechanism that does not require an objective message content to be legal. At the same time, the very survival of our economic system depends on the consumption increase, especially in recession periods like the current one (Díaz-Méndez 2010).

4.2 Sustainability

In modern, industrialised and urbanised societies, the complexities of the food system have long made fulfilling this obligation difficult and costly. Several factors have made carrying out this responsibility more serious and difficult in the present age. Government agencies responsible for food safety now face an evolving set of conditions that affect their abilities to fulfil their responsibilities. Among these new conditions are (1) the increasingly concentrated corporate control over the food chain from the farm to market, (2) a rapidly growing global market in even basic food staples and (3) hostile international political movements, which raise the threat of terrorist activities that could severely affect the food system. Each of these considerations is enough to make it necessary to have greater governmental oversight, regulation and inspection due to features of the urbanised society and the modern, industrial agricultural production system (Kaplan 2012).

Concern about the environment and sustainable growth has raised questions related to resource availability and limits regarding the ability of the planet to provide food for everyone with an improved material standard of living. Such concerns lead to charges that the industrialised world is living beyond its means and taking more than its share of resources to produce a lifestyle that is not sustainable. Moreover, they have profound implications for corporate activity that is based on the promotion of consumption and an ever-increasing material standard of living. Whether overconsumption is a legitimate problem and changing patterns of consumption are necessary are questions that need examination (Buchholz 1998).

There are many stakeholders in the food system and many stages between the farm and the dinner table. At each stage, problems can arise that may ultimately cause harm. From the perspective of rights theory, people have a right not to be harmed (or placed in harm's way). Utilitarian ethics demands that the greatest social good be pursued; this would imply that people should not be placed at risk without any overriding benefit. Either way, the possibility of harm, risk or hazard (however, one characterises a potential threat to life or health) establishes a moral responsibility on the part of "someone" to try to prevent or reduce that threat. It is commonly thought that "that someone" should be governments or their delegated authorities. To ensure, based on a validated system of risk analysis, that the food system is safe and secure is the primary ethical responsibility used to prevent harm. Though everyone may have a part in this collective responsibility, it is up to governments, acting on behalf of people's rights or the general welfare, to safeguard consumers and citizens as far as possible (Kaplan 2012).

Some 10 or so years ago, the term "responsible consumption" was coined to describe mindful consumption by concerned consumers who hold civic and social values. Responsible consumption implies that a consumer's buying decision is based on a set of criteria that may include the product's origin, manufacturing process, labour work conditions, environmental impact and the manufacturer's or dealer's social responsibility (Díaz-Méndez 2010). For such a process to function, consumers' education and information access are of utmost importance.

4.3   Food Policy

The food industry in developed countries has produced a huge variety of food that is not geographically or seasonally dependent. This food is so inexpensive that in countries like America, all but the very poorest can and do obtain enough energy and nutrients to meet their biological needs (Nestle 2007). On the other hand, availability of nutritious and safe food is very often assumed as being universal. This is not necessarily so, and indeed the United Nations has recognised the need to have it enshrined in the United Nations Declaration of Human Rights (UNDHR) which states, in Article 25, that "Everyone has the right to a standard of living adequate for the health and well-being of himself and of his family, including food, clothing, housing and medical care and necessary social services, and the right to security in the event of unemployment, sickness, disability, widowhood, old age or other lack of livelihood in circumstances beyond his control" (United Nations 1945).

Food is a necessity for life and causes problems only when consumed inappropriately. Promotion for the consumption of healthy food therefore becomes a complex issue if the primary mission of food companies is to sell products. The issue becomes further complicated when nutrition only becomes a factor in corporate thinking when it can be used to boost product sales (Nestle 2007). The ethics of such practices and corporate responsibilities associated with such a scenario are considered all too rarely.

To further confound the scenario, the emergence of global markets is known to have a significant impact on consumers' food selection, health, food security, social justice and quality of life. The whole of the food chain that covers production, distribution and retail all leading to consumption is in essence a global politics of food and health which is under scrutiny and tension.

The assumptions upon which this production is based, which are in the realm of experimentation in food policy, are termed Food Wars. There are three major paradigms of policy to be considered, namely, the prevailing arguments of the dominant industrial-protectionist model of the twentieth century versus the other two conflicting paradigms (one developing food through integrating the "life sciences", the other through "ecology"). These latter models compete to replace each other and the prevailing dominant productionist model. It is a war in which the interested parties are wrestling to attract investment, public support and significantly policy legitimacy over the appropriate use of biology and food technologies.

Food is an important component of our daily existence. It is not surprising therefore that it has shaped human development, influenced our past (history) and continues to exert its influence in everything today. Different visions for the future of food are shaping the potential for how food will be produced and marketed. Inevitably, there will be policy choices for the state, the corporate sector and the civil society (Heasman and Lang 2012). However, human and environmental health needs to be at the heart of these choices if healthy functional societies are to prevail.

Policy choices arise for varied reasons, and the transition from a paradigm in crisis to a new one from which a new science tradition can emerge is far from being an easy and natural growth process or one achieved by revamping old models. There are instances where it is the reformation of the field from new fundamentals, a process that changes some of the field's most elementary theoretical generalisations and fundamental tenets and applications. During a transition period, there will be a large but never complete overlap between the problems that can be solved by the old and by the new model. In addition, there will also be a decisive difference in the modes of solution. When the transition is complete, the community and profession will have changed its view of the field, its methods and its goals (Kuhn 2012). The Food War paradigm is no exception.

4.4 Eating and Food Production Tendencies

Early in the twentieth century, when the principal causes of death and disability were infectious diseases related in part to inadequate intake of calories and nutrients, the goals of health officials, nutritionists and the food industry in countries such as America were essentially the same and were intended to encourage people to eat more of all kinds of food (Nestle 2007).

In the food industry, the notion was to have optimization of production and the efficient use of raw materials to have as favourable a conversion from material to product as was possible. The emphasis here is on quantity, potentially at the expense of quality. The paradigm is "technology and supply" driven, with human health, if at all considered, an afterthought and where the consumer is merely a passive end user in the scheme. This view has dominated the twentieth century outlook to food production. Indeed, from the 1930s onwards, it developed a science base that supported the goals of increasing output. This model is referred to by Heaseman and Lang as the productionist paradigm (Heasman and Lang 2012).

Indeed, the Food and Agriculture Organization (FAO) in principle was an advocate for this model when it initially defined food security as the concept of "ensuring, to the utmost, the availability at all times of adequate world supplies of basic food stuffs, primarily cereals, to avoid acute food shortages in the event of widespread crop failures or national disasters, sustain a steady expansion of production and consumption, and reduce fluctuation in production and prices" (Carolan 2013).

Unfortunately, the ultimate goal of safe and health-enhancing food for all, produced in ways that do not adversely affect the environment and future generations, has not been achieved. The "productionist" vision measured success on continuously increasing output, where quantity delivered was the measure of success. Opinion is changing on this, and there is a school of thought that promotes the concept that the food system's efficacy has to be judged against more complex and broader criteria: environmental ones such as climate change, water, soil and correct use of land, together with health issues such as obesity, alongside hunger, escalating healthcare costs and safety. There is an emerging agenda in which issues such as public health and nutrition, social (widening inequalities within and between nations), cultural (marketing excess, perversion of needs), ethical (animal welfare, decent labour conditions) and economic (rapid concentration distorting markets, internalisation of full costs) are increasingly becoming important (Lang et al. 2009).

4.5   Choice: The Freedom to Consume Well

The freedom to follow one's own lifestyle of consumptive preferences is a core value in contemporary affluent societies. Any interventions have to be more than justified if they are to be acceptable to the society, especially if they are initiated by governments. With political liberalism becoming dominant in modern societies, the freedom to do as one wills, even in the consumption of products and services, is coming to the fore. This individual freedom is a function of pluralism in contemporary affluent and democratic societies as characterised by differing lifestyles and consumptive preferences, with individuals having their own take on what is good and acceptable. John Rawls states that "equal citizens have different and indeed incommensurable and irreconcilable conceptions of the good" (Rawls 1993; Beekman 2000). Political liberalism's emphasis on individual freedom is based on the assumption that consumers are above all motivated by the opportunity to follow their chosen lifestyles. People may disagree with the choice, but recognise the fact that all consumers should be equally free to follow the style that they have selected and that others should respect those choices. This respect for freedom to live one's lifestyle is part of the consensus model of contemporary affluent societies which value autonomy (Beekman 2000). The question to be asked is where and when do the concepts of autonomy and freedom from outside interference (including government) affect society and whether it is necessary to restrict this freedom for the sake of the greater good. This raises issues in the political liberalism model which argues that public entities such as governments should remain neutral (impartial) and not favour the arguments in favour or otherwise of lifestyle choice and consumptive preference.

4.5.1 Choice Based on Sustainability Arguments

If the growing demand for specific foods, such as dairy and meat, which is putting pressure on the environment were to be considered as an example of the need for change, it is hard to imagine that such a global concern can be solved solely by increasingly efficient technologies and agricultural practices. Lowering the consumption of those commodities that put undue pressure on the local and wider environment seems inevitable. Yet, the issue is whether modern consumers can be considered as reliable allies to do so to achieve this shift in consumption patterns, taking into consideration the tension between responsible intentions as citizens and the hedonic desires as consumers. Consumers can and should be considered as partners that must be involved in understanding the occurrence of alternative ways of commodities' consumption and thereby contribute to a more sustainable world.

FAO has observed that the global community will have to face significant challenges to feed 9 billion mouths in 2050 within the means of the global ecosystem (FAO 2009). The production and consumption of animal proteins are among the most environmentally harmful components of human eating preferences. The conversion of grains and other crops into animal foods is highly energy consuming making the consumption and production of animal proteins an urgent issue now more than ever. Experts predict enormous consequences for the environment, nature and landscape as well as food security if consumption patterns of animal proteins are not restricted and a more sustainable, plant-based diet promoted to replace it (UNEP 2007; FAO 2009).

With the presentation of such a scenario, there are those who argue that rigid views on moral responsibility versus irresponsible behaviour should be avoided and that consumer change should not only or primarily be associated with rational persuasion strategies. A realistic approach that questions the typical citizen-consumer tension has to be debated. Such an approach looks at various practices of sustainable food consumption rather than defining ethical or political consumption in desirable and strict terms. Ethical consumption when interpreted broadly within a framework identifies several routes of change. Such an approach would lead to more optimism about opportunities to find consumer involvement that may lead to the informed selection of sustainable commodities, including food (deBakker and Dagevos 2012).

4.5.2   Food Industry Influence on Food Choice

The marketing of food continues to evolve from its promotion as a tasty necessity for life to a potentially health-enhancing experience. Increased interest in the connections between food and health may have been driven by time-challenged individuals looking for better nutritional and efficient solutions that promote health. A survey carried out in 2000 revealed that over 50% of respondents were more likely to eat foods reported to reduce the risk of heart disease and cancer (Parker 2003).

Regulatory agencies and food marketers have responded to this interest with product labelling and advertising claims that inform consumers of vital information ranging from the product's fat content to claims that its consumption may decrease the likelihood of disease. In December 2006 the EU Council and Parliament adopted a harmonised regulatory framework on nutrition and health claims made on foods. For the first time, the use of nutrition claims such as "low fat" and "high fibre" or health claims such as "reducing blood cholesterol" was regulated across the EU. This regulation introduced measures to ensure that any claim made on food labelling, presentation or marketing in the European Union was clear, accurate and based on evidence accepted by the whole scientific community.

In the EU in 2011, the then Commissioner for Health and Consumers stated (in a memo) that citizens had the right to accurate and reliable information on food labels to help consumers make healthier choices (Statement by Commissioner John Dalli: Another step forward: more accuracy on health claims in food 2011). Furthermore, when health claims were involved, it was considered particularly important that the statements had to be both truthful and accurate. The 2006 Regulation EU No1169/2011 (implemented in 2011) on nutrition and health claims made on foods enabled such monitoring and controls. This so-termed Claims Regulation ensures high level of protection for consumers with clear labelling and any claims on the said labels having been fully justified (European_Commission 2013). As a first step, nutrition information on labels was changed as of December 2014 to provide clearer information regarding composition and declaration of potential content of allergens.

Nutrition labelling in the EU remains highly topical, and the recent restructuring provides greater transparency and provides information that favours the knowledgeable consumer (FoodDrink Europe 2013).

Popular books like *Fast Food Nation* (Schlosser 2001) and films like *Super Size Me* (Spurlock 2004) and Food, Inc. (Kenner and Pearlstein 2008) have sensitised the public to industry practices. In addition, the recent rise in popularity of various social media apps, such as Twitter, Facebook and LinkedIn, has brought otherwise little known snippets of information regarding food industry practices to the attention of anyone using a computer, tablet or smartphone and has also contributed to sensitise consumers. In turn, the industry has had to react to claims that it lures children into a lifestyle of unhealthy eating and infiltrates schools, buys loyalty from scientists and pressures administration officials into accepting weak and ineffective nutrition policies (Brownell and Warner 2009).

There are those who contend that markets corrupt children and infantilise adults and cause the Disneyization or McDonaldisation of society which cause consumers to lose any control they have reducing them to the status of passive victims of market strategies that prevent change to more sustainable consumption patterns. Consequently there are those who argue that consumers are in no way connected to sustainability (deBakker and Dagevos 2012).

There are those who contend that consumers are obstacles to food sustainability and are actually enemies of citizens and society. In a book called *Does Ethics Have a Chance in a World of Consumers?*, Zygmunt Bauman points out that unrestrained economic freedom and market criteria that reflect instant consumption, instant gratification and instant profit are highly damaging. If there is to be any hope, consumers have to consider the importance of others as an "ethical demand" on themselves, the other being the individuals or groups who have to bear the consequences of the consumers' actions. Therefore, the consumer, to be considered proper, has the ethical responsibility to consume with mindfulness towards others and society. This idea as a concept of concern for one's neighbour and community may not at all fit well in the setting of a consumerist utopia (Bauman 2009).

In the United States, personal responsibility has been invoked to protect the food industry from criticism, legislation and litigation. Legislation sponsored by Congressman Ric Keller (R-FL) in 2004 to ban lawsuits claiming health damages against fast-food restaurants is typical in emphasising personal behaviour: "We've got to get back to those old-fashioned principles of personal responsibility, of common sense, and get away from this new culture where everybody plays the victim and blames other people for their problems". The call that people should be better informed about the moral complications of their meat and other food consumption, and be urged to adopt a more sustainable lifestyle, seems like a voice crying in the wilderness of our supermarkets (deBakker and Dagevos 2012).

For a long time, the needs and demands of consumers were often manipulated for the sake of production. The Codex Alimentarius, the agency responsible for international food standards, provided the following definition for consumers: "persons and families purchasing and receiving food in order to meet their personal needs" (Codex_Alimentarius 2001). This definition, even if formulated to briefly characterise an individual, removes a considerable part of their identity, interest or responsibility in food safety. However, the 2011 EU regulation (EU No1169/2011) has given advocacy to consumers to have the freedom of choice and the right to know what is in the food they are buying. The more active role taken by consumers became especially significant whenever a food safety crisis occurred in Europe, and it was realised that close interconnections between economic interests and food policy can have negative consequences. In many European countries and elsewhere, this active consumer role arguably has been accompanied by a transfer of responsibility for food safety from the public sector to the individual (Gaivoronskaia and Hvinden 2006).

4.5.3   Choice as Person Identity

Humans are biologically adapted to their ancestral food environment in which foods were dispersed and expenditure of energy was required to secure them. With the evolution of culture and through the evolutionary process of preadaptation, now food serves functions other than nutrition. This change places the nutritional aspects of food in a broader and more complex context. Food develops into a social vehicle that allows people to make social distinctions and to establish social linkages. It is no longer just a means of sustenance; it also becomes a medium for aesthetic expression, giving rise to elaborate food preparations and cuisines that cannot be justified solely in terms of nutritional factors (Rozen 2005).

Food choice has increasingly become a form of expression of consumers' personality with the selection of some food types tending to reflect beliefs about valued ways of being or living and behaviours. "Life-guiding principles" interact with food choice motives (such as health, shopping or eating convenience, religious reasons, or ecological welfare) and constitute food ideologies that reflect the consumers' ideals and ways of living and also shape their food-related lifestyle (Bellows et al. 2010). People eat foods, not nutrients (or pure substance solutions), so a psychological taxonomy of foods would not at all resemble a nutritional classification (Rozen 2005).

4.5.4   Social and Cultural Influences on Consumerism

Quintessentially omnivores or generalists, humans do consider almost anything as a potential food, and through history there are various texts that have given meaning to everything connected with food: who is allowed to fish for it, farm it, kill it or mill it; what vessels and utensils are used in its preparation; what time of day the meal is eaten; and who sits where at the table (if a table is used at all).

Identity – religious, ethnic, national – is intensely associated with food. Every community thinks of itself as special and uses food to show it. Indeed, what defines a meal varies throughout the world. Meals are culturally prescribed eating bouts and often include multiple dishes whose order is often culturally prescribed. More than one dish may be available at any given point in a meal. In the United States and England, meat makes the meal. In ancient Rome, roast pig was the banquet centrepiece. In Mediaeval Europe, bread was the staple food, meaning that there was bread and often not much else. In Asia, the staple food is rice, and to be considered a meal, there must also be vegetables (Civitello 2011). Most people consume foods in the form of meals. If so, individuals have some options as to how to structure the time sequence in their own meals. Faced with a typical main course of two to four items, people may adopt varied eating patterns which may presumably generate different experienced and remembered pleasures of that event (Rozen 2005).

Consumption, in late twentieth-century western forms of capitalism, may be seen as a social and cultural process involving cultural signs and symbols, not simply as an economic, utilitarian process. Yet both in the advanced societies of capitalism, and in those social formations which remain predominantly rural and agriculturally based, there are many groups whose patterns of consumption are determined by their economic situation, rather than the social and cultural factors (Bocock 2002).

Nevertheless, once people have been influenced by what might be called the social and cultural practices associated with the ideology of modern consumerism, then even if they cannot afford to buy the goods portrayed in films, in the press and on television, they can and do desire them (Huxley 1958). Consumption is seen here, therefore, as being based increasingly upon desires, not simply upon need. In the social formations of western capitalism, however, the marketing of products through the use of signs and symbols in selling products to the majority of consumers has been linked to longings to reach perceived status, through consumption of objects such as clothes, or styles of furniture, and food itself. These desires do not disappear even in periods of economic recession and put stress on all affected whether or not they remain gainfully employed during depressed times. Identity becomes dependent on consumption of goods rather than productive work roles. The human sciences are linked logically, and manifestly, with philosophy, including epistemology (the branch of philosophy concerned with the nature and scope of knowledge), moral and political philosophy, philosophy of history and ontology (the description of the concepts and relationships that can exist for an agent or a community of agents). The human sciences cannot be detached from this philosophical underpinning as various forms of positivism[1] have sought to do (Bocock 2002).

4.5.5   Ethical Issues in Eating Meat

As a matter of strict logic, if an individual is opposed to inflicting distress on animals, but not to the painless killing of animals, that person may consistently eat animals who had lived free of all suffering and been instantly, painlessly slaughtered. The practical and psychological justification (of efficient slaughtering) may be more difficult to reach. If society considers the life of another being (an animal) merely a means of satiation and pleasure, then taste for a particular type of food is then no more than a means to an end, such that pigs, cattle and chickens are considered as things to be used. In such instances there may be no need to change their living conditions which however would most likely draw critical attention. Even if

[1] A philosophical system that holds that every rationally justifiable assertion can be scientifically verified or is capable of logical or mathematical proof and that therefore rejects metaphysics and theism. The New Oxford American Dictionary (Kindle Locations 644, 769–644, 770). Oxford University Press. Kindle Edition.

this were so, to most their eating habits are dear to those who hold them and are not easily altered. It is not practically possible to rear animals for food on a large scale without inflicting considerable suffering. Even if intensive methods are not used, traditional farming involves castration, separation of mother and young, breaking up social groups, branding and transportation to the slaughterhouse and finally slaughter itself. It is difficult to imagine how animals could be reared for food without these forms of distress. However, there is certainly no need to inflict unnecessary additional distress, a consequence of lack of professional farming processes, or more explicitly the infliction of pain through cruel, crude practices performed by some handlers; one has only to search the Internet and social media to see examples of such practices. Though some argue that this might be possibly done on a small scale, others contend that one could never feed today's huge urban populations with animals raised in stressless environments but yet raised for their meat. If at all, the meat so produced would be vastly more expensive than any meat currently produced, as rearing animals is already an expensive and inefficient way of producing protein. The meat of animals reared and killed with equal consideration for the welfare of animals while they were alive would be a delicacy available only to the rich (Pence 2001). There are those who argue that this would create inequalities and limitation to access. One such example is beef; there are gourmet speciality beef cuts, intended to attract the fairly affluent buyer, such as Aberdeen Angus beef, and even more so when one considers Kobe or Wagyu[2] beef from Japan which can garner exorbitant prices when compared to standard unbranded beef. However, the concept of meat quality and safety to human consumers of intensely reared animals has to be further investigated. Issues regarding the effect of feed quality and quantity and environmental stress on the health of the animal itself and then on the people who consume it have to be considered. Thus the economic issues alone are not the only factors that should be considered. The food production system has to be itself sustainable and in turn and more importantly must sustain and promote human health.

4.6 Food Product Formulation and Its Impact on Health

To have a food chain that produces safe, healthy and quality products, it is necessary to have an integrated approach to agriculture, food processing, food transport, sociology, nutrition and medicine to propose new approaches to food production with optimal human nutrition as a principle goal (Sands et al. 2009). Food not only ensures our survival but is a potential source of pleasure and general well-being. In order to survive, the human brain is required to optimise the resource allocation

---

[2] The Daily Mail (UK) on the 3 June 2014 stated that supplies of prized Wagyu beef, which originated in Japan and is renowned for its flavour and succulence, are normally sold in luxury stores, gourmet butchers and food websites for £100 per kilo – putting it out of reach for most families. See also http://www.dailymail.co.uk/news/article-2647838/Aldis-latest-lure-middle-classes-Prized-wagyu-beef-7-8oz-steak-just-quarter-standard-price.html (accessed on 13/07/2014).

such that rewards are pursued when relevant. This means that food intake follows a similar cyclical time course to other rewards with phases related to expectation, consumption and satiety. There is evidence that brain networks and mechanisms initiate, sustain and terminate the various phases of eating. Associated with this are the underlying reward mechanisms of wanting, liking and learning that lead to how human food intake is governed by both hedonic and homeostatic principles (Kringelbach et al. 2012). Hence, it is therefore necessary to ask whether it therefore is acceptable to formulate foods knowing that such formulations will interact with the brain's pleasure centres and will tempt the user to indulge and possibly overuse if not abuse the product.

The traditional fields of research have diverged into disparate and often poorly communicating scientific disciplines such as food science, plant genetics and breeding, plant pathology, soil science, nutrition, toxicology, food microbiology, epidemiology, human/animal/plant biochemistry and various fields of medicine. The goals of each discipline are largely compartmentalised, and most research, publication and funding mechanisms that ensue are fragmented in their outcomes; these trends clearly demonstrate that there is significant stratification and segregation in research (Heywood 2011). Global agricultural policy relating to human health has largely been directed at issues of food safety and, within this remit, has been reasonably successful. Yet it has not effectively addressed the far greater agricultural effects on health, namely, its contribution to the quality of human nutrition through the provision of sufficient quantities of the right nutrients. The end products of the production chain should therefore be tailored to the needs of the public. This is where future effort in aligning agricultural policy with health priorities needs to be focused and such efforts that will need to involve both governments and the private sector. Only by making health one of the drivers of agricultural policy will it be at all possible to feed a human population of 9 billion healthily in 2050 (Waage et al. 2011).

The relationship between agricultural food production and health is complex. Despite the fact that there is an abundance of food in the world, paradoxically hundreds of millions across the world face a range of nutritional problems. It has been reported that in 2009, "more than 1 billion people went undernourished as their food intake regularly provided less than minimum energy requirements, not because there was not enough food, but because people were too poor to buy it". In addition, many people who enjoyed a sufficient energy supply (measured in calories) suffered from poor or inadequate nutrition, either because they were unable to afford a healthy diet with essential micronutrients or that despite their affluence the quality of their diets was unhealthy due to excessive intakes of energy, largely from fat, added sugars and energy-dense processed foods (Butler 2010). A combination of such issues with lifestyles that are generally characterised by low levels of physical activity leads to problems of obesity and chronic diet-related illnesses such as diabetes and heart disease. Despite all the international attempts to address it, micronutrient malnutrition (MNM) remains a persistent global issue. MNM affects a greater proportion of the world's population than protein-energy malnutrition (Heywood 2011).

If global nutrition were enhanced, there would be significant potential to reduce acute and chronic disease, the need for healthcare and the cost of healthcare and increase educational attainment, economic productivity and the quality of life. Breeding nutritionally improved crops can play an important role in alleviating malnutrition given that the two most important drivers of food production are optimal yield and minimal cost. Unfortunately, enhanced *nutritional value* is not currently an important driver of most plant breeding efforts, and there are only a few well-known efforts to breed crops that are adapted to the needs of optimal human nutrition. Optimal human nutrition therefore needs to be factored into plant breeding that is heavily influenced by agronomic yield drivers.

It is well accepted that foods have changed at a considerably greater rate than inherent human physiology. The genes that construct and control human physiology have changed minimally in the last 10,000 years and essentially not at all in the past 40–100 years. Modern progress in hygiene and developments in medicine have led to reduced infant mortality and longer average life spans, especially in communities where cereal grain-based diets could be supplemented with other sources of nutrients. Nonetheless, many of the negative consequences of the transition to grain-based diets remain today (Sands et al. 2009).

Malnutrition handicaps its victims with fatigue, lethargy, susceptibility to pathogens and a range of health complications, thus limiting their ability to care for themselves, to obtain and benefit from education and to contribute to society.

Efforts to document the costs of malnutrition for human health and for society provide powerful testaments to the extent of these losses, e.g. malnutrition has multiple causes including lack of food (production), lack of resources to procure food (poverty and infrastructure) and impairment of ability to absorb nutrients (disease). When malnutrition is due to problems of food availability or procurement, it is also often associated with limited food diversity and nutritional quality. In fact, lack of access to a diversity of foods is an indicator of poverty, and conversely dietary diversity has been proposed as an indicator of food security (Sands et al. 2009).

4.6.1 Humanistic Nutrition: Food for Life

The current crucial challenge is to feed humans optimally so that suboptimal nutrition does not impair the functioning of human minds and bodies and prevent the attainment of full human potential. Though there have been initial efforts to improve the nutritional value of crops as food for humans, there is yet tremendous potential for significant further nutritional enhancement of crops. This form of enhancement would effectively exploit the application of basic research in genetics, plant physiology, biochemistry and the emerging new integrative field of systems biology. It remains, however, unlikely to occur as long as maximum and stable yield is the overriding driver of the economic competitiveness of plant varieties. This is especially important for peoples with access to a limited diversity of foods, be it for reasons of availability, culture or choice (Sands et al. 2009)·

4.6.2 *Brain Function and Food Intake*

In humans, it has been proposed that some of the cortical networks that evolved for the more advanced aspects of eating-related behaviour have been recycled and have come to trigger other higher cognitive functions. Human eating is not, however, governed solely by homeostatic processes (Sands et al. 2009). Pleasure and reward mechanisms play a central role in the control of human food intake. In addition, food intake is modulated by other factors such as genetics, circadian rhythms, reproductive status and social factors. Evidence for the complexity of eating behaviour can be found in the influence of all five primary sensory systems as well as the visceral sensory system and gut–brain interactions. Food intake is therefore driven by motivation and emotion which are, in turn, supported by reward and hedonic processing.

Reward consists of multiple subcomponents, including the dual aspect of hedonic impact and incentive salience. The former refers to the "liking" or pleasure related to the reward, e.g. the experiences of eating, whereas the latter refers to the "wanting" or desire to obtain the reward. Learning is an important component in decision-making regarding eating-related behaviour, where the brain must compare and evaluate the predicted reward value of various behaviours (Berridge and Robinson 1998). Ultimately, pleasure can be thought of as an important tool to control this balancing act between exploitation and exploration.

Saliva is the medium that bathes the taste receptors in the oral cavity and in which aroma and taste compounds are released when food is eaten and saliva contains enzymes and molecules that can interact with food. The interindividual variability of saliva composition in the population is not well known; and this variability could be linked to variability in food perception or liking (Neyraud et al. 2012). An area of nutritional research referred to as sensory analysis is specifically concerned with characterisation of people's likes and dislikes of certain foods and seeks to correlate these results with chemical composition and texture of foods.

4.6.3 *Food Reformulation*

Protein-containing foods are essential for the guarantee of adequate nutrition. The development of new sources of proteins and the optimization of the existing ones are issues of great interest and study. Moreover, the use of ingredients beneficial to health has been identified as a steadily growing trend in the food industry. In this way, functional foods refer to foods or food ingredients that provide specific physiological beneficial effects and/or reduce the risk of chronic disease beyond basic nutritional functions.

With the need to reformulate foods that are energetically less dense and more appropriate for less active lifestyles, most people currently have challenged the food industry to seriously contemplate reformulation of foods with ingredients that help to lower health risks, as in the case of substituting animal fats by vegetable fats and oils, obtaining foodstuffs low in cholesterol and saturated fats. Cheese is one such good example of a food matrix that could lend itself well to the incorporation of vegetable proteins (Rinaldoni et al. 2014). Furthermore, it should also be noted that the production of cheese analogues can be less costly than those products obtained only from animal proteins, thus making such a food more accessible to lower-income consumers. Of the vegetable proteins, soybean is a highly nutritious food material that contains well-balanced amino acids and desirable fatty acids, and it plays an important role as a protein source for many people around the word. There are various food formulations that incorporate soy proteins for various purposes, usually associated health benefits.

4.6.4   Food for Health

In the last decades, consumer demands in the field of food production have changed considerably. Consumers increasingly believe that foods contribute directly to their health. Today foods are not intended to only satisfy hunger but are expected to have attributes that are associated with the prevention of nutrition-related diseases and improve physical and mental well-being of the consumers. In this regard, functional foods play an outstanding role. The increasing demand for such foods can be explained by the increasing cost of healthcare, the steady increase in life expectancy and the desire, by older people, for improved quality of their later years.

Functional foods are found virtually in all food categories; however, products are not homogeneously scattered over all segments of the growing market. The development and commerce of these products is rather complex, expensive and risky, as special requirements have to be met. Besides potential technological obstacles, legislative aspects, as well as consumer demands, need to be taken into consideration when developing functional food. In particular, consumer acceptance has been recognised as a key factor to successfully negotiate market opportunities (Siro et al. 2008).

The term "functional food" was first used in Japan, in the 1980s, for food products fortified with special constituents that possess advantageous physiological effects. In most countries however, there is no legislative definition of the term, and drawing a borderline between conventional and functional foods is challenging even for nutrition and food experts. To date, a number of national authorities, academic bodies and the industry have proposed various definitions, ranging from the very simple to the more complex. Most definitions also suggest that a functional food should be, or look like, a traditional food and must be part of our normal diet.

A functional food can be targeted at the whole population or at particular groups, which may be defined, for example, by age or genetic constitution (European Union 2010). European legislation, however, does not consider functional foods as specific food categories, but rather a concept. Therefore, the rules to be applied are numerous and depend on the nature of the food and its intended use. The General Food

Law Regulation[3] is applicable to all foods. In addition, legislation on dietetic food, genetically modified organisms (GMOs), food supplements or on novel foods may also be applicable to functional foods again, depending on the nature of the product and intended use. Rather than regulating the product group per se, legislative efforts currently being developed are directed towards restricting the use of health claims on packages and in marketing (Siro et al. 2008).

4.7   Food Safety and Legislation

In all EU member states and many third countries, food safety and consumer protection are covered in national legislation. The EU has an all-encompassing legal instrument in Regulation (EC) No 178/2002 that was adopted by the Council and Parliament that collated the agreed general principles and requirements of Food Law.[4] This regulation was needed to provide a framework to ensure a coherent approach in the development of food legislation, with the final aim of reaching harmonisation that would not be a barrier to trade within the single market. It was designed to provide the general framework that promoted mutual recognition between member states (and third countries) for those areas not covered by specific harmonised rules but which were functioning within the internal market. This framework document covered definitions, principles and obligations encompassing the whole food chain to ensure "a high level of protection of human life and health, taking into account the protection of animal health and welfare, plant health and the environment" (Europa Summaries of EU Legislation 2011).

Current food safety policy is based on principles established or updated at the beginning of the 2000s in line with the integrated approach "from the farm to the fork". Transparency, risk analysis and prevention, the protection of consumer interests and the free circulation of safe and high-quality products within the internal market and with third countries form the very essence of these regulations. To ensure that error and misjudgement is minimised, the process involved what was termed the precautionary principle which had to be intrinsically featured in all food-related decisions. Though food is handled with what is often regarded as a dose of

---

[3] Regulation (EC) No 178/2002 of the European Parliament and of the Council of 28 January 2002 laying down the general principles and requirements of Food Law, establishing the European Food Safety Authority and laying down procedures in matters of food safety *OJ L 31,* 01/02/2002, p. 1–24.   http://eur-lex.europa.eu/legal-content/EN/ALL/;jsessionid=7STTT49LjWhQg9JP7352 4w595js3Q5RZzy9xyr4wldhscS3TnXhy!1318808511?uri=CELEX:32002R0178 (accessed 30th March 2014).

[4] Regulation (EC) No 178/2002 of the European Parliament and of the Council of 28 January 2002 laying down the general principles and requirements of Food Law, establishing the European Food Safety Authority and laying down procedures in matters of food safety *OJ L 31,* 01/02/2002, p. 1–24.   http://eur-lex.europa.eu/legal-content/EN/ALL/;jsessionid=7STTT49LjWhQg9JP7352 4w595js3Q5RZzy9xyr4wldhscS3TnXhy!1318808511?uri=CELEX:32002R0178 (accessed 30th March 2014).

over familiarity, safe and conservative approaches should be figured in food policy. The precautionary principle, detailed in Article 191 of the Treaty on the Functioning of the European Union (EU), features the implementation of rapid response procedures wherever there is the potential for danger to human, animal or plant health or to protect the environment, in particular, where scientific data do not permit a complete evaluation of the risk (Europa Summaries of EU Legislation 2011).

In most cases, the danger associated with a procedure or a product placed on the market, the burden of proof falls on European consumers and their representative associations except for medicines, pesticides and food additives. However, when an action is taken under the precautionary principle, the producer, manufacturer or importer may be required to prove the absence of danger. This possibility would be examined on a case-by-case basis. It cannot be extended generally to all products and procedures placed on the market. Its application is based on current knowledge and should involve evidence-based literature. This, therefore, begs the question whether and how food and its formulation should be monitored and what guidelines if any should be used to monitor foods to ensure safety.

The drafting and promulgation of legislation is difficult and complex. For example, the WHO recommends 400 g of fruit and vegetable per day which approximates to five portions per day (WHO 2003). There is general consensus that eating vegetable is an important component of a healthy diet if consumed in sufficient amounts, with potential beneficial health effects as demonstrated in observational studies describing the relationship between fruits, vegetables and health. Yet concerns have been raised related to the presence of inter alia anti-nutrients (e.g. leptins, glucosinolates), nitrate, allergens, mycotoxins, contaminants and pesticide residues in vegetables. Leafy vegetables, in particular (e.g. lettuce, rocket salad, spinach), contain naturally high levels of nitrate (NO_3^-) which, together with its metabolites nitrite (NO_2^-) and nitric oxide (NO), all have not only physiological and potential beneficial roles in the body but also potential for adverse effects (Bottex et al. 2008).

4.8 Conclusions

The challenges that have to be addressed in the twenty-first century are therefore complex. While health promotion experts exalt the benefits of healthy eating and publish the results of the consequences of not doing so, there has to be a policy that enables communities as a whole to be able to afford to adopt such practices. If the cost of unhealthy eating to national health schemes were given more consideration, governments might find it worth their while to delve deeper into the real cost of less than adequate food choices that certain cohorts within communities are forced to adopt.

Food chain issues in the twenty-first century are complicated by globalisation and the prevailing economic landscapes. At one end of the spectrum, there are the large corporations which justify their activities as providers of the foods that are requested by consumers. At the other end of the spectrum, there are the consumers who are either unmanageable or consume in a hedonistic fashion. In this milieu, there are those who seek to feed themselves but are unable to do so or do so badly for reasons that range from limitations to access finance and time.

In the past, obesity[5] (especially morbid obesity) and overweight[6] together with comorbidities such as diabetes and vascular disease were observed only in adults. This is no longer the case with children being increasingly affected. Diabetes is a very expensive disease to treat. If rising numbers are not somehow curtailed and mitigated, the diabetes epidemic could break the bank of any healthcare system in the world, however funded. Effective solutions are therefore needed and soon. If people will not eat different foods, the foods people do eat have to be analysed and the potential for reformulation seriously considered to improve their nutrient profiles. It is worth noting that in developed countries, processed products account for 70–85% of intakes. The nutritional quality of these products therefore has a very significant influence on the nutritional status of these nations, for ill or for good (Winkler 2014).

The issues regarding rights and obligations of those individuals operating in the prevailing nutritional scene would need positioning in society. The healthcare, food science and associated professions do have opinion on what needs to be addressed to promote and produce healthy foods and have an ethical if not moral obligation to advocate for such foods. The issue may be the means needed to implement their advice. To change a consumer's favourite product to a healthier formulation may be fraught with issues which range from perception to taste and texture. Hence, the rights of the consumer as a chooser come in play as choice lies at the centre of the idea of consumerism. Indeed, such choices offer numerous advantages including a greater variety that is good for consumers; choice is good for the economy and is a driving force for efficiency, growth and diversity; a social system based on choice is better; consumer capitalism means more choice for all (Gabriel and Lang 1999).

Yet it should be emphasised that choice without information is not real choice. Taking the example of GM foods, it should be noted that industries have an aversion to positive labelling, and due to the apparent ineffectiveness of non-label methods of providing the information, very often consumers cannot make an informed choice about GM or any other products without a mandatory labelling policy (Streiffer and Rubel 2003). This may also be the case with other components found in foods. The contention starts with what sort of information should be given and to what level of detail, issues regarding marginal choice between products and issues when choice is available only to those with the adequate finances (Gabriel and Lang 1999). Yet issues may also arise when too much choice is available, as consumers may be confused by all the jargon presented to them. Indeed, choice can also be used as a means of offering variety and shedding responsibility or even worse be a source of deception.

---

[5] An adult who has a body mass index of 30 or higher is considered obese (source http://www.cdc.gov/obesity/adult/defining.html accessed 15/07/2014).

[6] An adult who has a body mass index between 25 and 29.9 is considered overweight. (Source http://www.cdc.gov/obesity/adult/defining.html accessed 15/07/2014).

There are those who contend that the advent of corporate social responsibility (CSR) has led to industry influencing consumers to differentiate between products and organisations using CSR as a push mechanism in the supply chain. Given the capability to form opinion (through knowledge) and therefore select in an informed fashion, consumers are exercising consumer social responsibility (C_NSR) and ultimately are the arbiters of behaviour.[7] The capability for choice available to the final beneficiaries of products on the market has the potential to shape markets and ultimately the products that the manufacturers produce. C_NSR has power because the citizens/consumers make a deliberate choice based on personal and moral beliefs which ultimately leads to the consumer engagement with the products and services (Manning 2014). Hence, C_NSR plays an important role in the consumerism versus ethical elements of acquisition. The degree of influence is not uniform across all consumers or (food) products but surely will influence choice which in turn will affect food chain composition and management.

References

Bauman Z (2009) Does ethics have a chance in a world of consumers? Harvard University Press, Cambridge

Beekman V (2000) You are what you eat: meat, novel protein foods and consumptive freedom. J Agric Environ Ethics 12(2):185–196

Bellows AC, Alcaraz GV, Hallman WK (2010) Gender and food, a study of attitudes in the USA towards organic, local, U.S. grown, and GM-free foods. Appetite 55:540–550

Berridge K, Robinson T (1998) What is the role of dopamine in reward: hedonicimpact, reward, learning, or incentive salience? Brain Res Rev 28:309–369

Bocock R (2002) Consumption (key ideas), Kindle Edition. Taylor and Francis, London and New york

Bottex B, Dorne J-LC, Carlanfer D, Benford D, Przyrembel H, Heppner C, Cockburb A (2008) Risk-Benefit health assessment of food – food fortification and nitrate in vegetables. Trends Food Sci Technol 19:S113–S119

Brillat-Savarin J-A (2012) The physiology of taste: or, meditations on transcendental gastronomy (1826; Washington, DC: counterpoint, 2000) in the rhetoric of food: discourse, materiality, and power (Routledge studies in rhetoric and communication). In: The rhetoric of food: discourse, materiality and power, Kindle edn. Taylor and Francis, London, UK

Brownell K, Warner KE (2009) The perils of ignoring history: big tobacco played dirty and millions died. How similar is big food. Milbank Q 87(1):259–294

Buchholz RA (1998) The ethics of consumption activities: a future paradigm? J Bus Ethics 17(8):871–882

Butler D (2010) Food: the growing problem. Nature 466:546–547. Retrieved 7 13, 2014, from Nature International weekly journal of science: http://www.nature.com/news/2010/100728/full/466546a.html

Carolan M (2013) Reclaiming food security. Taylor and Francis, London

Civitello L (2011) Cuisine and culture: a history of food and people, Kindle Edition. Wiley, Hoboken

Codex Alimentarius (2001) Codex Alimentarius, 1985 (revised 2001). Codex general standard for the labelling of pre-packaged foods. Rome. Retrieved 10 29, 2013, from ftp://ftp.fao.org/docrep/fao/005/y2770E/y2770E00.pdf

---

[7] See Chap. 6 for definitions of CSR and C_NSR.

deBakker E, Dagevos H (2012) Reducing meat consumption in today's consumer society: questioning the citizen-consumer gap. J Agric Environ Ethics 25:877–894

Díaz-Méndez M (2010) Ethics and consumption: a difficult balance. Int Rev Public Nonprofit Mark 7:1–10

Europa Summaries of EU Legislation (2011) The precautionary principle, Brussels. Retrieved January 5, 2014, from http://europa.eu/legislation_summaries/food_safety/general_provisions/l32042_en.htm

European Union (2010) Functional foods. Publications Office of the European Union, Luxembourg. doi:10.2777/85512

European_Commission (2013) Health and consumers, food. Retrieved October 29, 2013, from EUROPA: http://ec.europa.eu/food/foodlabellingnutrition/claims/

FAO (2009) How to feed teh world in 2050. FAO. Retrieved October 05, 2015, from http://www.fao.org/fileadmin/templates/wsfs/docs/expert_paper/How_to_Feed_the_World_in_2050.pdf

FoodDrink Europe (2013) Guidance on the provision of food information to consumers regulation (EU) No1169/2011. Food Drink Europe, Bruxelles

Frye JJ, Bruber MS (2012) The rhetoric of food: discourse, materiality, and power (Routledge studies in rhetoric and communication). Routlege Taylor and Francis Group, New York

Gabriel Y, Lang T (1999) The consumer as a chooser. In: Gabriel Y, Lang T (eds) The unmanageable consumer. Sage, London, pp 27–46

Gabriel Y, Lang T (2008) New faces and new masks of today's consumer. J Consum Cult 8(3):321–342

Gaivoronskaia G, Hvinden B (2006) Consumers with allergic reaction to food: perception of and response to food risk in general and genetically modified food in particular. Sci Technol Hum Values 31(6):706–710

Guptill AE, Copelton DA, Lucal B (2013) Food and society: principles and paradoxes, Kindle Edition. Wiley, London

Heasman M, Lang T (2012) Food wars: the global battle for mouths, minds and markets. Taylor & Francis, London

Heywood H (2011) Vernon, ethnopharmacology, food production, nutrition and biodiversity conservation: towards a sustainable future for indigenous peoples. J Ethnopharmacol 137:1–15

Huxley, A (1958) Brave new world revisited part VI the art of selling. Retrieved December 29, 2013, from http://www.huxley.net/bnw-revisited/

Kaplan DM (2012) The philosophy of food. University of California Press, Berkley

Karl Marx: The eighteenth brumaire of Louis Bonaparte. (n.d..) Retrieved April 12, 2014, from Liberty. Equality, fraternity-exploring the French revolution http://chnm.gmu.edu/revolution/d/580/

Kenner R, Pearlstein E (2008) Food Inc [motion picture]. Magnolia Pictures, Los Angeles

Kringelbach ML, Stein A, van Hartevelt TJ (2012) The functional human neuroanatomy of food pleasure cycles. Physiol Behav 106:307–316

Kuhn T (2012) The structure of scientific revolution: 50th anniversary ed. University of Chicago, Chicago

Lang T, Barling D, Caraher M (2009) Food policy integrating health, environment and society. Oxford University Press, Oxford

Manning L (2014) Consumer and corporate social responsibility in the food supply chain. Food Sci Technol 28(1):42–43

Nestle M (2007) Food politics: how the food industry influences nutrition and health. University of California Press, Berkeley and Los Angeles

Neyraud E, Palicki O, Schwartz C, Nicklaus S, Feron G (2012) Variability of human saliva composition. Possible relationships with fat perception and liking. Arch Oral Biol 57(5):556–566

Parker BJ (2003) The use of nutrient content, health and structure/function claims in food advertising. J Advert 32(3):47–55

Pence GE (2001) The ethics of food: a reader for the twenty-first century. Rowman & Littlefield Publishers, Oxford, UK

Rawls J (1993) Political liberalism. University Press, New York/Columbia. (cited by Beekman 2000)

Rinaldoni ANRPD, Noemi Z, Campderrós ME (2014) Soft cheese-like product development enriched with soy protein concentrates. LWT Food Sci Technol 55:139–147

Rozen P (2005) The meaning of food in our lives: a cross-cultural perspective on eating and well-being. J Nutr Educ Behav 37:S107–S112

Sands D, Morris C, Dratz E, Pilgeram A (2009) Elevating optimal human nutrition to a central goal of plant breeding and production of plant based foods. Plant Sci 177:377–389

Schlosser E (2001) Fast food nation: what the all-american meal is doing to the world. Penguin Books, New York

Siro I, Kapolna Emese KB, Lugasi A (2008) Functional food. Product development, marketing and consumer acceptance. A review. Appetite 51:456–467

Spurlock M (2004) Supersize me. Motion Picture, USA

Statement by Commissioner John Dalli: another step forward: more accuracy on health claims in food, MEMO/11/869 (2011)

Streiffer R, Rubel A (2003) Choice versus autonomy in the GM food labeling debate: comment. AgroBioForum 6(3):141–142. Retrieved March 30, 2014, from http://www.agbioforum.org

UNEP (2007) Global environment outlook GEO4: environment for development. Progress Press/ United Nations Environment Program, Valletta

United Nations (1945) United Nations declaration of human rights article 25, 1945. Retrieved 10 23, 2013, from United Nations: http://www.un.org/en/documents/udhr/index.shtml#a25

Waage J, Dangour A, Hawkesworth S, Johnston D, Lock K, Poole N, Uauy R (2011) WP1: understanding and improving the relationship between agriculture and health. http://www.bis.gov.uk/assets/bispart. Retrieved October 24, 2013, from Foresight Project on Global Food and Farming Futures: http://www.bis.gov.uk/assets/bispartners/foresight/docs/foodandfarming/additional-reviews/11-597wp1-understanding-improvingagriculture-and-health.pdf

WHO (2003) WHO technical report series 916. WHO, Geneva

Winkler J (2014) Nutritional reformulation: the unobtrusive strategy. Food Sci Technol 28:37–40

Winson A (1994) The intimate commodity: food and the development of the agro-industrial complex in Canada. University of Toronto Press, Toronto. Retrieved April 09, 2014, from http://www.amazon.ca/The-Intimate-Commodity-Development-Agro-Industrial/dp/0920059198

Chapter 5
Ethical Issues in the Food Supply Chain

Judith Schrempf-Stirling

5.1 Introduction

The food supply chain encompasses a wide range of businesses and activities. Broadly speaking this chain includes the following: agriculture and farming, food processing and manufacturing, food engineering, food transportation and distribution, food marketing, retailing, and restaurants.

There are various ethical issues in the food industry starting with the sourcing of the different ingredients over the actual production of our food and packaging to marketing and selling the product to the end consumer.

Ethical issues in the food industry go beyond the sale of food products. While food satisfies one of our most substantial needs to survive, it can equally contribute to health concerns. Thus, further ethical issues arise when it comes to the negative consequences of food consumption, such as food poisoning, allergic reactions, and weight gain. There have been numerous publications around different ethical issues in the food industry. Documentaries such as *Food Inc.*, *We feed the World*, or *Our Daily Bread* or books such as *Fast-Food Nation* (Schlosser 2002), *The Omnivore's Dilemma: A Natural History of Four Meals* (Pollan 2006), or *Food Politics* (Nestle 2013) provide insights into the food supply chain and some of the major ethical concerns such as farming standards, genetically modified food, lobbying, or food poisoning. Other documentaries focus on specific issues in the food industry, such as *King Corn* which focuses on one of the main ingredients in our food – corn – or *Food Chains* which provides inside views of the bad working conditions on farms.

In the first part of this chapter, I provide an overview of the most prevalent production-related issues in the food industry such as working conditions, environ-

J. Schrempf-Stirling (✉)
Robins School of Business, University of Richmond, Richmond, VA, USA
e-mail: judith.stirling@richmond.edu

© Springer International Publishing AG 2018
R. Costa, P. Pittia (eds.), *Food Ethics Education*, Integrating Food Science and Engineering Knowledge Into the Food Chain 13,
https://doi.org/10.1007/978-3-319-64738-8_5

mental issues, and engineering food. In the second part of this chapter, I provide an overview of some prominent consumption-related issues in the food industry such as marketing practices and obesity.

5.2 Production-Related Ethical Issues in the Food Industry

With the transformation and industrialization of the food industry did not only come advantages such as efficient production and employment. Numerous practices along the food supply chain have negative impacts on the environment, people, and health. In the following, I provide a brief overview of three common ethical issues in the food supply chain: working conditions, environmental impact, and engineering food.

5.2.1 Working in the Food Supply Chain

While the actual work in the food supply chain ranges from growing crops, raising animals, packaging meat, and serving meals in a restaurant, the working conditions along the food supply chain are very similar. Whether employees work on a farm, in a factory, in a grocery store, or in a restaurant, most of them are subject to low wages, health and safety risks, and less than perfect working conditions.

Did you know that one farm worker dies on the job each day in the United States (McCluskey et al. 2013)? A recent report summarized the severe working conditions under which farm workers suffer. For instance, wages are so low that farm workers live below the poverty line. Besides, the working and living conditions are dangerous and unhealthy which explains the abovementioned death statistic. Heat is one of the main issues in agriculture and farming. Farm workers are exposed to the sun and heat the whole day. Besides, classic farm work consists of a high level of physical work including heavy lifting: "Heavy loads and repetitive motion strain workers' bodies. Slips, trips, and falls happen on a regular basis. Irrigation equipment can electrocute workers. Tractors overturn. Workers can become entrapped in grain silos and engulfed in clouds of pesticides. In short, farm work is dangerous business" (McCluskey et al. 2013).

In the next stage of the food supply chain, working conditions are not much better. Food manufacturing – like farming – provides a great job opportunity for those with low educational background or immigrants. Studies show that over 50% of workers in food manufacturing are immigrants (Spangher 2014). Despite this upside, there is a downside to food manufacturing. News, articles, and reports describe the hazardous working conditions, and workers say that they feel like "a piece of trash" (Genoways 2014) and describe how they defecate in their pants because they must not leave their working station or slow down the process (Spangher 2014). Food manufacturing workers suffer from low pay, long working hours, and dangerous working conditions that are linked to the fast pace of the factoring line (Spangher 2014).

Reports describe how employees do not get adequate health and safety training and as a consequence are injured on the job. Food manufacturing workers describe how they are struck by equipment or fall on wet floors. In a New Yorker tortilla factory, one worker died because of a missing guard on a flour-mixing machine. Some studies estimate that the number of workers injured on the job reaches 40% (Brandworkers and the Community Development Project 2013).

Other issues in food manufacturing include low wages, discrimination, lack of unionization, and unpaid overtime. Workers in food manufacturing earn almost $8 less than the industry average. Besides, wages differ significantly between female and male workers. Male workers earn more and are more likely to move faster to higher positions than their female counterparts. There is not only discrimination based on gender but also based on nationality. According to Brandworkers and the Community Development Project (2013), undocumented immigrants earn less than immigrants of legal status. Besides, many workers complain about unpaid overtime and facing retaliation when they try to organize a union to increase their bargaining power.

Finally, there are numerous long-term negative effects on workers' health. For instance, workers in food manufacturing suffer significantly more from repetitive motion injury than the average worker. Repetitive motion injury occurs when workers engage in the same type of motions over a long period of time. They use the same muscles, tendons, or joints, which might eventually lead to fatigue, discomfort, or permanent posture issues.

Finally, working conditions in retailing and restaurants – one of the last stages in the food supply chain – are not much better. A recent study summarized the wage level and its consequence on employees of restaurant chains such as McDonalds or Burger King (Allegretto et al. 2013). According to this study, employees earn around $9 per hour or only the federal minimum wage of $7.25. Given the employees' family situation, this wage level leaves them below the poverty line, and most of them rely on state and federal support programs such as food stamps. Net earnings of some of the restaurant chains reach over $1.5 billion per quarter. However, their employees do not earn enough to make their own living. The low price, low cost business model in food retailing has had similar devastating effects on its employees. A recent study described the decline of wages and working conditions in food retailing in California (Jayaraman and the Food Labor Research Center 2014). More than half of the employees do not earn enough to meet the low standard of living. Jayaraman and the Food Labor Research Center (2014) highlighted that working conditions and wage levels are significantly better in retail stores that are unionized. Workers in retail stores with a union earn on average $3 more compared to their counterparts in non-unionized retail stores.

One might assume that the working conditions of employees in the food industry is better in countries with more protective labor laws such as Germany or other European countries. However, working conditions for employees in the food industry are still far from perfect even in such protective environments. Wage levels in restaurant chains are similarly low in the United Kingdom compared to the United States (Royle 2005). Besides, other worker rights are violated. There have been cases where employees were encouraged to engage in off-the-clock work which basically means to work without pay. Examples include working while being on a break or working before or after the official shift starts (Royle 2005).

Thus, no matter where in the world, the working conditions in the food industry are less than ideal.

5.2.2 Environmental Footprint of the Food Industry

Did you know that 40% of our land surface is used to provide the world population with food? Did you know that 30% of our land surface grows grains, fruits, and vegetables to feed to animals (chicken, pigs, and cattle) that eventually end up on our plate? The upside of these numbers is that farming and livestock production employs over 1 billion people worldwide (Herrero et al. 2013), but the downside is that our demand for food comes at an environmental cost. Our food affects the environment on each of its production stages:

- Farming and harvesting
- Storage and transportation
- Food packaging and food waste

5.2.2.1 Farming and Harvesting

The agricultural side of the food chain, that is farming and harvesting, relies heavily on natural resources and the ecosystem. Farmers need soil, water, and countless other organisms to grow crops and raise animals. However, as discussed in the following, many farming practices have a negative impact on our environment.

Farmers rely heavily on chemicals to provide crops with nutrients (fertilizers) and to protect their crops from pests (pesticides). While these chemicals aim at supporting food production, they also contribute to environmental pollution and have negative effects on the ecosystem. For example, fertilizers reach the groundwater or can be carried to close waterways such as local water streams and rivers and pollute the water. Fertilizers, for example, can deplete oxygen from water and thereby kill organisms and animals living in the water (Diaz and Rosenberg 2008; Howarth 2008).

Pesticides have similar negative effects. Application of pesticides has been linked to soil fertility degradation and deformities in amphibians (Mulvaney et al. 2009). Scientists even believe that pesticides might have caused the recent decline in bee population (Klein et al. 2007; Horrigan et al. 2002; Spivak et al. 2011). A further problem with pesticides is that insects and plants eventually develop pesticide resistance which forces farmers to apply *more* concentrated pesticides or higher quantities of pesticides to achieve the desired protection effect, thereby intensifying the negative consequences of pesticide usage on the environment and ecosystem.

Farming depends on a number of natural resources and is considered being a major contributor to the heavy decline of the very same resources. For instance, farmers need fertile soil, water, fossil fuels, and different kinds of minerals such as phosphate which is used in the production of chemical fertilizers. Unfortunately, these resources are used at a higher rate than they are reproduced. Thus, some of our fundamental resources such as water are declining and becoming scarce. For example, two-thirds of freshwater is used for irrigation in the United States – most of this is used in agriculture (John Hopkins Center for Livable Future 2014). The recent 2015 water crisis in California illustrates the reality of water shortage. The food sector has been criticized for using and wasting high amounts of water and also for its potential to pollute groundwater or river streams through the use of fertilizers and pesticides as mentioned earlier (Horrigan et al. 2002; Hart et al. 2004).

Given these negative consequences on the environment and ecosystem, alternative approaches to farming, such as organic farming, have been considered.

Organic farming does not use synthetic pesticides or petroleum-based fertilizers. Besides, organically raised animals have access to the outdoors, are fed organic feed, and are free of hormones (USDA Agricultural Marketing Service 2008). According to a Swedish study, raising organic beef produces 40% less greenhouse gas emissions and consumes over 80% less energy compared to conventional farming practices (Estrada-Flores 2008). While organic farming constitutes a great alternative, it is not without criticism. Federal guidelines and standards about what constitutes organic practices have been criticized as not being strict enough and allowing the use of certain pesticides or fertilizers that are still harmful to the environment. Besides, some of the big food companies have entered the organic market so that some organic farms actually look more like large industry-scale farms supplying distant supermarket chains. Supporters of organic farming criticize this industrialization trend as they perceive the organic market as representing a small-scale local production including local distribution. According to them there is no room for big business.

Overall, there are various environmental issues when it comes to growing our food. Environmental issues also exist at the next stage of the supply chain: storage and transportation.

5.2.2.2   Storage and Transportation

Food transportation is essential in the food supply chain and happens at every stage – from farms, to production centers, to warehouses, to export destinations, supermarkets, and finally to the consumers' home (Estrada-Flores 2008). Thus, it is not surprising that food transportation constitutes almost a fourth of all the petroleum used globally and is estimated to cause more than 10% of carbon emissions from fossil fuels (Estrada-Flores 2008). "Food miles" evaluate the "environmental friendliness of foods on the basis of the transportation distances required to bring products to consumers from the farm/grower" (Estrada-Flores 2008). Thus, the lower the food miles, the more environmentally friendly is the product. In short, food miles capture the environmental footprint of food transportation.

However, the distance is not the only aspect of transportation that has an impact on the environment. Cooling is an essential part of food transportation. Cooling is required to keep our food fresh. However, cooling requires energy. In Australia alone it has been estimated that the refrigeration-related process of cooling, freezing, air conditioning, and cold storing of food uses 20,000 gigawatt hours per year which is equivalent to 18 megatonnes of CO2e.

CO2e refers to equivalent carbon dioxide and measures carbon footprint. It represents the amount of CO2 that would cause the same amount of global warming than any greenhouse gas (The Guardian 2011). To put this number further into perspective, the energy used in refrigeration processes in Australia per year is equivalent to that of 4.3 million cars on the roads (Estrada-Flores 2008).

5.2.2.3 Food Packaging and Food Waste

A further component that has an environmental impact is food packaging. Food packaging does not only refer to the final packaging or containers that consumers buy in supermarkets. Food packaging also includes secondary packaging such as the larger boxes that our products come in for transportation. Finally, packaging also includes any other packaging material that is used during transportation such as wooden pallets or plastic wrapping (Estrada-Flores 2008). It is estimated that Europe causes almost 80 million tons of packaging waste, while packaging constitutes up to 14% of solid waste in New Zealand. While more than 50% of package waste is recycled worldwide, it still leaves us and our planet with a significant amount of waste.

One sad consequence of this packaging waste is that some of it ends up in our ocean. A recent study (McDonnell 2015) estimated that there are between 4.8 and 12.7 million metric tons of plastic in our oceans. One example of this is the Great Pacific Garbage Patch in the North Pacific Ocean. The effects on the ocean and its habitants from plankton to fish and whales are devastating. Birds, fish, and other sea animals die from eating the plastic, or they get tangled in it while passing. A recent study investigated the main source of the plastic in the ocean and concluded that the top plastic polluters are middle-income countries with a growing coastal population lacking proper waste management. China, India, and the Philippines belong to the top plastic polluters (McDonnell 2015).

This raises an important point when discussing the environmental impact of the food supply chain: the role of the consumer. Consumers play a role in how they treat product waste such as plastic bottles, food cans, or other food packages. Besides, consumers create waste at the end of the food life cycle. A 2012 consumer report estimated that American consumers throw out almost a quarter of their food and beverages. A British study puts the percentage of wasted food even higher – at 33%. The monetary value of wasted food in British households lies at around £8 billion each year (Estrada-Flores 2008). Given all the food miles, the carbon footprint, and energy used to produce these products, food waste is a double burden. The produced food has had already environmental impacts and is now thrown away, creating issues of disposal and wasting consumers' money (Gunders 2012). Bloom (2011) estimates that the average American family of four wastes $1364 each year due to food waste. Food waste does not only happen at consumers' home. It also appears at all different stages in the food supply chain. For example, it is estimated that US supermarkets, restaurants, and convenience stores discard spoiled food that is worth $30 billion every year (Estrada-Flores 2008). Thus, even at the end of the food life cycle, ethical issues persist.

5.2.3 Engineering Food

Food corporations are criticized for engineering food in ways that verge on exploitation of human psychological and physical drives (Moss 2013). For instance, critics claim that food producers use consumers' natural addiction to sugar, salt, and fat to their advantage (Stein et al. 2012). Even though food companies add certain ingredients such as sugar, salt, and caffeine for flavoring, such ingredients also make consumers want to have more (Moss 2013). While food companies claim that the added caffeine enhances the flavor of products such as soft drinks, energy drinks, potato chips, jellybeans, or candy bars (Brownell and Warner 2009), most consumers are not able to detect caffeine taste in such products. Thus, critics (Griffiths and Vernotica 2000) conclude that "the high rates of consumption of caffeinated soft drinks more likely reflect the mood-altering and physical dependence-producing effects of caffeine as a central nervous system-active drug than its subtle effects as a flavoring agent."

What is even more troublesome is the optimization of food ingredients (especially the mixture of salt, sugar, and fat) to reach the so-called bliss point – "the precise amount of sweetness – no more, no less – that makes food and drink most enjoyable" (Moss 2013: 47). This bliss point evokes almost the same satisfaction received from some recreational drugs (Moss 2013) and can also undermine consumers' feeling of satiety: food melts so quickly in the mouth that the brain does not notice any calories. Hence, consumers can eat more. Popcorn is a classic example of a food with vanishing caloric density.

Thus, engineering the optimal mix of ingredients so that consumers do not develop a sense of satiety and might even feel "high" from consuming food raises some serious concerns. There appears to be a potential for consumers to become addictive to certain types of products or ingredient constellations. Humans have a natural tendency and preference for sugar as a source of quick energy, and research has shown that humans can become addicted to substances such as sugar, caffeine, salt, or fat (Moss 2013). We cannot blame anyone for natural addictions. However, we might feel differently if someone uses such urges to their own advantages. Food corporations engage in engineering food to design their products in such a way to increase the addiction potential so that consumers buy more (Moss 2013). Critics highlight that there has never been such high levels of salt in our food than today and that "the manufacturers of processed foods have been creating a desire for salt where none existed before" (Moss 2013).

While some might argue that we should make use of science and optimize food taste where we can, others see a fundamental issue with engineering food: from an ethical perspective, it is wrong to exploit physical, cognitive, motivational, or social vulnerabilities of consumers for someone's own advantage when such vulnerabilities are outside consumers' control (and often awareness) (Brenkert 1998).

5.3 Consumption-Related Issues in the Food Industry

The previous section provided a brief overview of some ethical concerns in regard to the production of our food. The consumption side of food also raises ethical issues ranging from deceptive marketing practices to the contribution of obesity – one of the biggest health concerns of the twenty-first century.

5.3.1 Marketing

Marketing has been described as one of the least ethical aspects of the corporation (Baumhart 1961). Thus, it is not surprising that the food industry is criticized for its marketing efforts. Common allegations include that food companies provide misleading nutrition information, knowingly deceive the public, and play a major role in the rise of obesity. The marketing practices of the food industry in general and that of the fast-food sector in particular have been described as deceptive and manipulative. Adams (2005) criticizes food companies because they "promote the sale of food high in sugar, fat and sodium content, unfairly target vulnerable consumers, encourage overconsumption, fail to provide patrons with the information needed to make informed decisions at the point and time of purchase, and ultimately shift or externalize the costs associated with consumption".

What makes advertisement in the food sector challenging is that some types of food might not be good for consumers. Some food groups such as sweets or junk food – food that is high in fat and sugar – are considered as harmful and critics question whether it is ethical to advertise such products to consumers in the first place. Research, for example, found that the harmfulness of the product has a high effect on the ethical evaluation of marketing practices (Smith and Cooper-Martin 1997). Thus, the actual product qualities (or the lack thereof) play a big role in ethical judgment. Hence, one of the most obvious ethical issues in the food industry is marketing: is it right to market a product that is high in fat and sugar and that might lead to negative health consequences such as diabetes?

Corporations in general often dispose of advertisement budgets of several billion dollars. Fast-food corporations alone spend more than $4 billion on advertising per year. This includes TV, radio, the Internet, billboards, as well as social media. For food companies, the majority of their marketing focuses on the promotion of processed food and food that is high in fat and sugar. Fast-food companies, for instance, advertise mainly their standard products and promote their healthier food options such as salads significantly less (Harris et al. 2010). Food advertisement is effective. A study on preschool and school children found that children were seeing between three and five fast-food TV spots daily. As a consequence, children asked their parents regularly (sometimes even daily) to visit a fast-food restaurant to eat the advertised products (Harris et al. 2010). This links to the second marketing challenge in the food industry: marketing to children.

Food companies are criticized for unfairly targeting children (Adams 2005; Barboza 2003; Seiders and Petty 2004). The challenge with marketing to children is twofold: children might not be aware that they are being marketed and children are easy to influence.

From a business perspective, it makes sense to market to young consumers early. The earlier a consumer uses a product, the earlier that consumer creates brand loyalty and habit. Obviously, food advertisement has an effect on children because children want to eat what they see and what is specifically marketed to them (Strasburger and Wilson 2002). Cartoon characters are a popular marketing tool in the food industry. Research has shown that branding food packages with licensed characters positively influences children's taste and snack preferences. The effect of cartoon characters is fascinating. In one study, researchers (Roberto et al. 2010) found that 78% of the participating children chose a chocolate bar over broccoli when none of the food packages had cartoon characters. However, when the researchers put an Elmo sticker on the broccoli package, half of the children chose the broccoli over chocolate (Roberto et al. 2010). This experiment shows two things: First, cartoons can be a powerful tool in attracting children. Second, cartoons might have a stronger effect on making children eat a certain product than the product itself. We would have expected that children would always pick chocolate over broccoli (with or without any cartoon character on either product). However, it shows that a cartoon character can be enough to persuade children to eat something healthy despite having the option to eat something tasty like chocolate.

In addition to directly targeting children in advertisement, food companies apply other practices to attract children. Restaurants of fast-food giant McDonald's, for example, have a colorful playground for children. McDonald's has a bigger playground than private entities in the United States (Schlosser 2002). Besides, fast-food restaurants offer special kid's menus that include toys. Toys are often linked to current popular kids' movies for recognition value and to increase popularity for such toys among children. Also, fast-food restaurants offer a selection of toys which triggers a desire to collect the different toys – thus, encouraging follow-up visits and purchases. Critics claim that the kid's menus in fast-food restaurants do not meet nutritional needs and are too high in fat, sugar, and salt. In response to that critique, San Francisco introduced a new regulation and set nutritional requirements for meals that are sold with toys (CSPI 2010). Such requirements include that the kid's meal has less than 600 calories and includes a certain amount of fruit and vegetables. Fast-food companies found a legal loophole and responded to the regulation by disconnecting the toy from the kid's meal. McDonald's and Burger King now offer a toy for 10 cents with the purchase of their kid's menu. Thus, the main objective of the regulation – improving the nutritional value of kid's meals at fast-food restaurants – has apparently failed.

Also other governments have reacted to the concerns in regard to food advertisement to children. Norway, for example, restricts advertisement of unhealthy food to children under 16 (Johnston 2012). In a similar vein Ireland forbids TV advertisement of sweet and fast-food products. Also, companies are not allowed to use celebrities to endorse their products in case of junk food when targeting young consumers (Flanagan 2013). Likewise, the US Federal Trade Commission has developed voluntary guidelines regarding junk food marketing to children (Nestle 2011). The guidelines suggest that the advertised food include at least one healthful food group such as vegetables or fruits and restrict the quantity of less healthful ingredients such as fat, sodium, or sugar.

5.3.2   Price Promotions

The second element in the marketing mix is price. Two ethical issues arise when looking at pricing in the food industry. First, healthier food options appear to be more expensive. Organic food, for example, is priced at a premium. On the flipside of that, junk food and other less healthy food options are more affordable. The biggest challenge here is that consumers in lower-income groups might not be able to afford healthier food. This creates social justice issues. The documentary *Food Inc.* showcases this concern when it shows a family of four who often eats fast food because it is cheap. In one scene the family visits a supermarket, and it becomes obvious that it is impossible to find a full meal at a comparable price.

The second concern, when it comes to pricing, is price promotion. Price promotions are used to increase the immediate awareness of the promoted food, triggering spontaneous positive consumer reaction and signaling to consumers the popularity of the promoted product (Hoek and Gendall 2006; Naylor et al. 2006). Research has shown that price promotions affect consumers' preference for the promoted product and signal to consumers that the product is frequently eaten (Hoek and Gendall 2006) – why else would this product be on sale? Thus, price promotions signal product popularity and encourage repeated purchases. Being frequently exposed to food advertisements, influences consumers' attitudes, and makes them believe that the consumption of the promoted food is normal (Grier et al. 2007).

Hoek and Gendall (2006) go as far as arguing that "food advertising is not innocuous and that, in depicting frequent consumption and consumption of larger portion sizes ("up-sizing") as normal behaviors, it contributes to the rise in obesity."

5.3.3   Labeling

Food and menu labels are defined as a "display of standardized information about the nutrient content of food and beverages in a readily available, clearly visible format at the point of sale, intended to inform people's purchasing and consumption decisions" (Mah and Timmings 2015). Food labels can take different forms, such as a table highlighting and listing the amount of food ingredients or a classification sticker "low fat." Food and menu labels can be a source of information and provide consumers with an opportunity to make healthy food choices. However, food labels can also raise ethical concerns because they can be misleading.

Food corporations are denounced for deceiving consumers about the true ingredients through misleading labels. When food companies were criticized for excessive amounts of sugar in their products, Kellogg renamed some of its leading brands: "Sugar Frosted Flakes" became "Frosted Flakes" and "Sugar Smacks" to "Honey Smacks" (Moss 2013). The word sugar was exchanged with "honey" and "fruit" which still did not change the amount of sugar. Thus, the sugar amount remained untouched, but the name change disguises one of the most prominent ingredients to the consumer (Moss 2013).

Wansink and Huckabee (2005) provide an example of the effects of misleading food labels such as "low fat." In their experiment, more than a third of restaurant consumers believed that they were eating less calories than they actually did. These misperceptions were due to in-store ads and healthy food labels (Wansink and Huckabee 2005). Consumers often assume that "low fat" products are healthier without reading the ingredient list properly. Some low-fat products are indeed low in fat but high in sugar. As this effect is known in research, it might also be known in business. Thus, this creates an ethical issue. Do food companies misuse consumers' reliance on labels?

To counteract the negative effects of food labeling, some governments have started to introduce official labels. In the United Kingdom, the government has launched a food traffic light labeling to provide consumers with better guidance in regard to the healthiness of food (Food Standards Agency 2007). The traffic light box is at the front of the food package and highlights the amount of fat, saturates, calories, salt, and sugar. A red light signals high, yellow light medium, and green light low levels of fat, saturates, calories, salt, and sugar. Thus, the food traffic light system allows consumers to immediately see how healthy the food is (the more green lights, the healthier the food). However, it took a considerable long time until the food traffic light was introduced. Germany and the European Union discussed the introduction of a similar labeling system but failed to do so. The food industry strongly opposed such labeling propositions.

Finland provides a promising example of how effective proper food labeling can be. The country had a significant high rate of cardiovascular disease which is linked to high consumption of salt. Thus, the Finnish government introduced food labels on products that had high levels of salt. The labeling was successful: consumers bought less salty food, and food companies started reducing the amount of salt in the products (Moss 2013). Thus, this example showcases the famous win-win scenario: consumers paid attention to the food label, and corporations responded to the decline in consumer demand by adapting their products.

5.3.4   Obesity

In 2010 the World Health Organization referred to obesity as an epidemic, and in 2013 the American Medical Association declared obesity a disease. Obesity refers to high levels of body fat or adipose tissue in relation to lean body mass. An estimated 600 million people are obese globally. Despite efforts to reduce obesity rates, obesity is increasing worldwide. The increase is visible in both developed countries such as the United States or Great Britain and developing countries such as Mexico. Obesity is related so several severe health issues and illnesses such as increased risks of stroke, coronary heart diseases, and type 2 diabetes. Experts disagree on what exactly causes obesity because weight gain is influenced by many factors such as genetic, metabolic, behavioral, environmental, socioeconomic, and sociocultural factors (Malik et al. 2006; Nielsen and Popkin 2003). However, the food industry is on top of the list as one cause of rising rates of obesity.

"The foods that are offered in restaurants, snack shops, and vending machines are higher in sugar, calories, and fat than what we typically prepare in our own homes. We are surrounded by food. We're constantly bombarded by it. We're consuming larger portion sizes and more calories than ever before" (Centers of Disease Control and Prevention 2011: 1).

Such criticism against the food industry is not that absurd if one considers the effect of food ingredients such as fat and sugar, portion sizes, and the seduction to overconsume.

Processed food refers to any type of food that has been deliberately changed. Such processing can be as simple as freezing or drying food, but it can also include many complex steps. Processed food normally includes chemical ingredients such as additives or artificial flavorings. A rule of thumb is that if you cannot find the food product as is in nature, then it is processed. Also, a long list of ingredients can give consumers an indication of the level of processing. Examples of processed foods are fruit and vegetables cans, packaged foods such as cereals, poultry, and food served in fast-food restaurants to name a few.

Researchers criticize that some types of processed foods such as fast food are very energy dense. That is, they provide a high level of calories in a relatively small amount of food. Thus, consumers might consume more calories. For example, energy density in fast food or highly processed food can be up to 65% higher than the average diet (Stender et al. 2007). Harris et al. (2010) found that only 0.5% of kid menu items in major fast-food restaurants meet proper nutrition criteria for children. In general, only 17% of all items in fast-food restaurants qualify as healthy (but most of those are beverages) (Harris et al. 2010). Consequently, consumers increase their daily caloric intake by 100–700 calories with one visit at a fast-food restaurant (Harris et al. 2010).

It comes therefore at no surprise that energy-dense processed foods as well as sugar-sweetened beverages are linked to weight gain and obesity (Basu et al. 2012; Malik and Hu 2006; Moreno and Rodrigues 2007; Hu and Malik 2010). Thus, processed food might constitute a health risk, and nutritionists recommend to consume certain types of processed foods such as fast food in very small amounts only (Ludwig et al. 2001; Ludwig and Rostler 2007). Some governments have started to take action when it comes to processed food. Mexico, for example, introduced an 80% tax on junk food which they defined as high-calorie foods with 275 calories or more per 100 grams of food. Foods that meet this criterion are potato chips, candy, pudding, and peanut butter (Cohen 2013).

A further factor that plays a role in increased caloric intake when consuming food is the portion size. Portion sizes at restaurants have increased over the past decades (Young and Nestle 2002). Besides, food packages such as chips, chocolate, and other products are relatively generous. Larger portion sizes encourage consumers to eat more. The food amount on a plate signals to the consumer a normal or appropriate size to consume (Birch et al. 1987; Fisher et al. 2003). This can have a dangerous effect and lead to overeating. Wansink et al. (2005) set up an experiment and compared how much food participants would eat. One group of participants received a normal-sized bowl with soup, while the second group of participants received a self-refilling bowl with soup. The participants were unaware of the self-refilling bowl. The researchers found that participants in the second group consumed over 70% more compared to the first group with a normal bowl. Besides, participants with the self-refilling bowl underestimated the amount of consumed calories and stated that they did not feel significantly more saturated than participants in the first group with the normal bowl. Thus, the portion size is a critical component when it comes to overeating. Consequently, one way to control portion sizes could be to encourage smaller packaging.

A final ethical issue in regard to the food industry's role in the rise of obesity has less to do with the product or portion size but rather with the industry's communication around obesity. Weight gain is a result of more energy intake than usage – we eat more than we burn through exercise. Thus, there are two components to obesity: a poor diet – high in processed food – and lack of exercise. Medical experts assign a greater role to a poor diet for causing obesity. However, in their communications food companies keep stressing the lack of exercise as the cause of obesity. For instance, in an interview PepsiCo CEO Indra Nooyi said: "If all consumers exercised, did what they had to do, the problem of obesity wouldn't exist" (Mangalindan 2010). This overemphasis on lack of exercise by the food industry is referred to as "leanwashing" (Karnani et al. 2014).

In general, food companies refer mainly to the energy intake and burning equation signaling consumer responsibility for obesity (Kentucky Fried Chicken 2008b; McDonald's Corporation 2006; Pizza Hut 2008). Thus, the individual should make reasonable food choices and engage in enough physical activity to avoid a calorie imbalance. Lawyers of food companies underline the reasoning of consumer responsibility by arguing that "every responsible person understands what is in products such as hamburgers and fries, as well as the consequences to one's waistline, and potentially to one's health, of excessively eating those foods over a prolonged period of time" (Santora 2002). Hence, the one who decides about eating a hamburger must be the one who is responsible.

Critics claim that the food industry consciously deflects from its own role in obesity by focusing on the lack of exercising and thereby externalizing the responsibility for obesity. The leanwashing strategy by the food industry contributes to misinforming the public about potential causes of weight gain and obesity. This is why Karnani et al. (2014) conclude: "by systematically deflecting the public discourse from bad diet to exercise and other factors, the food industry is at least partly responsible [...] and is thus culpable in perpetuating the obesity epidemic."

5.4   Conclusions

This chapter provided a brief overview of some of the most prevalent ethical issues in regard to the production and consumption of food. Most of these issues are relevant in both developed and developing countries and apply to a wide range of consumers.

I want to close this chapter by focusing on ethical challenges for a particular consumer group: low-income consumers.

When it comes to food, the challenge for consumers of lower socioeconomic status is threefold: First, their budget is restricted, and they are thus not able to afford all kinds of food. Processed foods, such as fast food and soda drinks, are more affordable than healthier food options. Studies in France show that many healthy food items such as fruit and vegetables or low-fat meat are almost 50% more expensive than comparable less healthy food items (Temple and Steyn 2009). Monsivais and Drewnowski (2007) provide some numbers and argue that healthy foods can cost as much as ten times more than high-calorie, less healthy food. An African consumer described her food choice situation as follows: "I'm going to go towards whatever is on sale even if that's the worst thing for me. I have to eat somehow and still be able to pay my phone bill, cable, all this other stuff" (Antin and Hunt 2012). Low-income consumers' budget is restricted, and they might simply not be able to buy healthy food. A study in low-income neighborhoods in Los Angeles and Sacramento illustrated that 70% of the budget of a low-income household would be needed to meet the Dietary Guidelines for Americans when it comes to fruit and vegetables alone (Cassady et al. 2007). That amount would not leave much room for other essential monthly expenditures such as rent, water, and electricity – let alone other types of food products such as meat. Some studies have shown that there is a positive correlation between obesity (a result of a bad diet) and income level (Pacheco Santos 2013; Kac et al. 2012; Basile et al. 2012). Thus, the pricing of food products raises social justice issues and puts more burden on low-income consumers.

A second challenge is access to healthy food. Many low-income neighborhoods are described as "food deserts" – an area without direct access to fresh and healthy food. Thus, these neighborhoods oftentimes lack grocery stores, markets, or other forms of healthy food providers (U.S. Department of Agriculture 2014).

Food deserts have often convenience stores. However, convenience stores mainly offer processed food and food that is high in fat and sugar. Most convenience stores do not sell healthy food options such as fresh fruit and vegetables (Farley et al. 2009; Bodor et al. 2010). It is therefore not surprising that research found a positive relationship between obesity rates and the proximity to convenience stores (Morland et al. 2006; Powell et al. 2007). Thus, the existence of food deserts in mainly low-income neighborhoods raises social justice issues, and food retailers have been called upon their social responsibility to open supermarket outlets in food deserts (Sims and Felton 2006). Entering impoverished areas, though, constitutes a financial risk to supermarket chains, as the community does not dispose of a lot of spending power. Howlett et al.(2016) highlight the positive effect that supermarkets can have in entering food deserts: their study shows that obesity rates decrease with an increase in grocery stores and supermarkets because consumers get more access to healthier food options. With the establishment of a supermarket, there can be an increase in access to fruit and vegetables of up to 30% (Morland et al. 2006). Thus, the question is whether food retailers have a responsibility to serve all communities equally.

Finally, consumers of lower socioeconomic status are generally less educated and lack knowledge in regard to food and how food choice relates to their health. Thus, this consumer group might not be aware of the importance of food choices and consequences to their health. Considering these disadvantages raises some critical questions when it comes to certain marketing efforts of food companies. For instance, research has found that more promotion (in-store advertisement) of less healthful menu options has been found in poorer African-American neighborhoods than in affluent white areas (Lewis et al. 2005). Also, Block et al. (2004) found that there were significantly more fast-food restaurants per square mile in predominantly African-American neighborhoods in Louisiana compared to predominantly white neighborhoods. Thus, while many of the issues discussed in this chapter affect all kinds of consumers, some consumer groups might be exposed to additional ethical issues – raising concerns of discrimination.

References

Adams RJ (2005) Fast food, obesity, and tort reform: an examination of industry responsibility for public health. Bus Soc Rev 110(3):297–320

Allegretto SA, Doussard M, Jacobs K, Thompson D, Thompson J (2013) Fast food, poverty wages: the public cost of low-wage jobs in the fast-food industry. Berkeley Center for Labor Research and Education, Berkeley

Antin TM, Hunt G (2012) Food choice as a multidimensional experience. A qualitative study with young African American women. Appetite 58(3):856–863

Barboza D (2003) If you pitch it, they will eat it. New York Times. Retrieved Sept 12, 2009, from http://www.nytimes.com/2003/08/03/business/if-you-pitch-it-they-will-eat.html.

Basile N, Paniqui S, Tarico L, Moratal I (2012) Anthropometric diagnosis of weight and height and feeding strategies of a vulnerable population. Diaeta 30(140):11–17

Basu S, Stuckler D, Galea G, McKee M (2012) Nutritional determinants of worldwide diabetes: an econometric study of food markets and diabetes prevalence in 173 countries. Public Health Nutr 13:1–8

Baumhart RC (1961) How ethical are businessmen? Harv Bus Rev 39(4):6–176

Birch LL, McPhee L, Shoba BC, Steinberg L, Krehbiel R (1987) Clean up your plate: effects of child feeding practices on conditioning of meal size. Learn Motiv 18:307–317

Block JP, Scribner RA, DeSalvo KB (2004) Fast food, race/ethnicity, and income: a geographic analysis. Am J Prev Med 27(3):211–217

Bloom J (2011) American wasteland: how America throws away nearly half of its food (and what we can do about it). Da Capo Press, Cambridge

Bodor JN, Rice JC, Farley TA, Swalm CM, Rose D (2010) The association between obesity and urban food environments. J Urban Health 87:771–781

Brandworkers and the Community Development Project (2013) Feeding New York. Challenges and Opportunities for Workers in New York City's Food Manufacturing Industry. Urban Justice Center, New York

Brenkert G (1998) Marketing and the vulnerable. Bus ethics quart. Ruffin Series 1:7–20

Brownell KD, Warner KE (2009) The perils of ignoring history: big tobacco played dirty and millions died. How similar is big food? Milbank Q 87(1):259–294

Cassady D, Jetter KM, Culp J (2007) Is price a barrier to eating more fruits and vegetables for low-income families? J Am Diet Assoc 107(11):1909–1915

Centers of Disease Control and Prevention (2011) The obesity epidemic. Retrieved Oct 10, 2011 from http://www.cdc.gov/CDCTV/ObesityEpidemic/Transcripts/ObesityEpidemic.pdf

Cohen, L (2013) New soda tax makes Mexico a leading guardian of public health. Huffington Post. Retrieved June 15, 2015 from http://www.huffingtonpost.com/larry/new-soda-tax-makes-mexico_b_4325724.html

CSPI 2010. San Francisco Moves Closer to Fast-Food Toy Marketing Curbs. Available online: https://cspinet.org/article/san-francisco-moves-closer-fast-food-toy-marketing-curbs, accessed August 1, 2017

Diaz RJ, Rosenberg R (2008) Spreading dead zones and consequences for marine ecosystems. Science 321(5891):926–929

Estrada-Flores S (2008) The environmental dimension of food supply chains chain of thought a newsletter of food chain. Intelligence 1(2):18

Farley TA, Rice J, Bodor JN, Cohen DA, Bluthenthal RN, Rose D (2009) Measuring the food environment: shelf space of fruits, vegetables, and snack foods in stores. J Urban Health 86(5):672–682

Fisher JO, Rolls BJ, Birch LL (2003) Children's bite size and intake of an entree are greater with large portions than with age-appropriate or self-selected portions. Am J Clin Nutr 77:1164–1170

Flanagan, P (2013) Junk food adverts to be banned from children's television. Irish Mirror. Retrieved June 13, 2015 from http://www.irishmirror.ie/news/irish-news/junk-food-adverts-banned-childrens-1931533

Food Standards Agency (2007) Food. Using traffic lights to make healthier choices. Retrieved Jan 7, 2010 from http://www.food.gov.uk/multimedia/pdfs/publication/foodtrafficlight1107.pdf

Genoways T (2014) 'I felt like a piece of trash' – life inside America's food processing plants. The Guardian. Retrieved June 20, 2015 from http://www.theguardian.com/world/2014/dec/21/life-inside-america-food-processing-plants-cheap-meat

Grier SA, Mensinger J, Huang SH, Kumanyika SK, Stettler N (2007) Fast-food marketing and children's fast-food consumption: exploring parents' influences on an ethnically diverse sample. J Pub Policy Mark 26:221–235

Griffiths RR, Vernotica EM (2000) Is caffeine a floring agent in cola soft drinks? Archiv Fam Med 9:727–734

Gunders D (2012) Wasted: How America is losing up to 40 percent of its food from farm to fork to landfill. Natural Resources Defense Council Issue Paper. Retrieved from http://www.indianasna.org/content/indianasna/documents/NRDC_Wasted_Food_Report.pdf, June 23, 2015

Harris J L, Schwartz M B, and Brownell K D (2010) Fast food f.a.c.t.s. Food Advertising to Children and Teens Score. Yale Rudd Center for Food Policy and Obesity, Yale

Hart MR, Quin BF, Nguyen M (2004) Phosphorus runoff from agricultural land and direct fertilizer effects. J Environ Qual 33(6):1954–1972

Herrero M, Havlík P, Valin H, Notenbaert A, Rufino MC, Thornton PK, Obersteiner M (2013) Biomass use, production, feed efficiencies, and greenhouse gas emissions from global livestock systems. Proc Natl Acad Sci 110(52):20888–20893

Hoek J, Gendall P (2006) Advertising and obesity: a behavioral perspective. J Health Commun 11:409–423

Horrigan L, Walker P, Lawrence RS (2002) How sustainable agriculture can address the environmental and public health harms of industrial agriculture. Environ Health Perspect 110(5):445

Howarth R (2008) Coastal nitrogen pollution: a review of sources and trends globally and regionally. Harmful Algae 8(1):14–20

Howlett E, Davis C, Burton S (2016) From food desert to food oasis: the potential influence of food retailers on childhood obesity rates. J Bus Ethics 139(2):215–224

Hu FB, Malik V (2010) Sugar-sweetened beverages and risk of obesity and type 2 diabetes: epidemiologic evidence. Physiol Behav 100:47–54

Jayaraman S, The Food Labor Research Center (2014) Shelved. How wages and working conditions for California's food retail workers have declined as the industry has thrived. United Food and Commercial Workers, Western States Council. Berkeley, CA

John Hopkins Center for a Livable Future (2014) Agriculture and ecosystems. Retrieved from http://www.jhsph.edu/research/centers-and-institutes/teaching-the-food-system/curriculum/_pdf/Agriculture_and_Ecosystems-Background.pdf, Accessed 23 June 2015

Johnston, A (2012) Sorry Norway Food Companies, No More Greasy Burger Ads for Kids. The Epoch Times. Retrieved June 12, 2015 from http://www.theepochtimes.com/n2/world/sorry-norway-food-companies-no-more-greasy-burger-ads-for-kids-323868.html

Kac G, Velasquez-Melendez G, Schlüssel MM, Segall-Côrrea AM, Silva AA, Pérez-Escamilla R (2012) Severe food insecurity is associated with obesity among Brazilian adolescent females. Public Health Nutr 15(10):1854–1860

Karnani A, McFerran B, Mukhopadhyay A (2014) Leanwashing: a hidden factor in the obesity crisis. Calif Manage Rev 56(4):5–30

Kentucky Fried Chicken (2008) Nutrition. Retrieved from http://www.kfc.com/nutrition/, Oct 13, 2008

Klein AM, Vaissiere BE, Cane JH, Steffan-Dewenter I, Cunningham SA, Kremen C, Tscharntke T (2007) Importance of pollinators in changing landscapes for world crops. Proc Royal Soc London B: Biol Sci 274:303–313

Lewis LB, Sloane DC, Nascimento LM et al (2005) African Americans' access to healthy food options in South Los Angeles restaurants. Am J Public Health 95(4):668–673

Ludwig DS, Peterson KE, Gortmaker SL (2001) Relation between consumption of sugar-sweetened drinks and childhood obesity: a prospective, observational analysis. Lancet 357:505–508

Ludwig DS, Rostler S (2007) Ending the food fight: guide your child to a healthy weight in a fast food/fake food world. Houghton Mifflin, Boston

Mah CL, Timmings C (2015) Equity in public health ethics: the case of menu labelling policy at the local level. Public Health Ethics 8(1):85–89

Malik VSM, Hu FB (2006) Intake of sugar-sweetened beverages and weight gain: a systematic review. Am J Clin Nutr 84:274–288

Mangalindan J P (2010) PepsiCo CEO: 'If all consumers exercised...obesity wouldn't exist'. Fortune. Retrieved from http://archive.fortune.com/2010/04/27/news/companies/indra_nooyi_pepsico.fortune/index.htm, June 15, 2015

McDonald's Corporation (2006) 2006 Worldwide Corporate Responsibility Report

McCluskey, M, McGarity, T, Shapiro S, and Shudtz, M (2013) At the Company's mercy: protecting contingent workers from unsafe working conditions. Center for Progressive Reform.

McDonnell T (2015) There's a Horrifying Amount of Plastic in the Ocean. This Chart Shows Who's to Blame. Retrieved from http://www.motherjones.com/environment/2015/02/ocean-plastic-waste-china, June 22, 2015

Monsivais P, Drewnowski A (2007) The rising cost of low-energy-density foods. J Am Diet Assoc 107(12):2071–2076

Moreno LA, Rodríguez G (2007) Dietary risk factors for development of childhood obesity. Curr Opin Clin Nutr Metab Care 10:336–341

Morland K, Diez Roux AV, Wing S (2006) Supermarkets, other food stores, and obesity: the atherosclerosis risk in communities study. Am J Prev Med 30(4):333–339

Moss M (2013) Salt sugar fat: how the food giants hooked us. Random House, New York

Mulvaney RL, Khan SA, Ellsworth TR (2009) Synthetic nitrogen fertilizers deplete soil nitrogen: a global dilemma for sustainable cereal production. J Environ Qual 38:2295–2314

Naylor RW, Raghunathan R, Ramanathan S (2006) Promotions spontaneously induce a positive evaluative response. J Consum Psychol 16(3):295–305

Nestle M (2013) Food politics: how the food industry influences nutrition and health, vol 3. University of California Press, Oakland, CA

Nestle M (2011) New federal guidelines regulate junk food ads for kids. The Atlantic. retrieved from http://www.theatlantic.com/health/archive/2011/04/new-federal-guidelines-regulate-junk-food-ads-for-kids/238053/, June 23, 2015

Nielsen SJ, Popkin BM (2003) Patterns and trends in food portion sizes, 1977–1998. JAMA 289(4):450–453

Pizza Hut. (2008) Nutrition information. Retrieved from http://www.pizzahut.com/Nutrition.aspx, Aug 3, 2008

Pollan M (2006) The omnivore's dilemma: a natural history of four meals. Penguin, New York

Powell LM, Auld MC, Chaloupka FJ, O'Malley PM, Johnston LD (2007) Associations between access to food stores and adolescent body mass index. Am J Prev Med 33(4):S301–S307

Roberto CA, Baik J, Harris JL, Brownell KD (2010) Influence of licensed characters on children's taste and snack preferences. Pediatrics 126(1):88–93

Royle T (2005) Realism or idealism? Corporate social responsibility and the employee stakeholder in the global fast-food industry. Bus Ethics A European Rev 14(1):42–55

Santora M (2002) Teenagers' suit says McDonald's made them obese. New York Times 21(B):1

Santos LMP (2013) Obesity, poverty, and food insecurity in Brazilian males and females. Cad Saude Publica 29(2):237–239

Schlosser E (2002) Fast food nation. Houghton Mifflin, New York

Seiders K, Petty RD (2004) Obesity and the role of food marketing: a policy analysis of issues and remedies. J Publ Policy Mark 23(2):153–169

Sims RR, Felton EL Jr (2006) Designing and delivering business ethics teaching and learning. J Bus Ethics 63(3):297–312

Smith NC, Cooper-Martin E (1997) Ethics and target marketing: the role of product harm and consumer vulnerability. J Market 61(3):1–20

Spangher L (2014) The overlooked plight of factory farm workers. The Huffington Post. retrieved June 20, 2015 from http://www.huffingtonpost.com/lucas-spangher/plight-of-factory-farm-workers_b_5662261.html

Spivak M, Mader E, Vaughan M, Euliss NH Jr (2011). The plight of the bees. Environ Sci Technol 45 (1), pp 34–38

Stein LJ, Cowart BJ, Beauchamp GK (2012) The development of salty taste acceptance is related to dietary experience in human infants: a prospective study. Am J Clin Nutr 95(1):123–129

Stender S, Dyerberg J, Astrup A (2007) Fast food: unfriendly and unhealthy. Int J Obes 31:887–890

Strasburger VC, Wilson BJ (2002) Children, adolescents and the media. Sage Publications, Thousand Oaks

Temple NJ, Steyn NP (2009) Food prices and energy density as barriers to healthy food patterns in cape town, South Africa. J Hunger Environ Nutr 4(2):203–213

The Guardian (2011) What are CO2e and global warming potential (GWP)? Retrieved June 23, 2015 from http://www.theguardian.com/environment/2011/apr/27/co2e-global-warming-potential

USDA Agricultural Marketing Service (2008) National Organic Program

U.S. Department of Agriculture (2014) Food deserts. Agricultural Marketing Service. retrieved June 14, 2015 from http://apps.ams.usda.gov/fooddeserts/foodDeserts.aspx

Wansink B, Huckabee M (2005) De-Marketing obesity. Calif Manage Rev 47(4):6–18

Wansink B, Painter JE, North J (2005) Bottomless bowls: why visual cues of portion size may influence intake. Obes Res 13(1):93–100

Young LR, Nestle M (2002) The contribution of expanding portion sizes to the US obesity epidemic. Am J Public Health 92(2):246–249

Part III
Ethics for Food Professionals

Chapter 6
Codes of Ethics of Food Professionals: Principles and Examples

Paola Pittia

6.1   Introduction

Food has an essential role in our everyday life, by fulfilling basic nutritional human needs. Quality and safety are essential properties of the foods, and by the centuries, actions and technologies have been developed and improved to guarantee health and well-being of humans worldwide.

The food manufacturing sector, including all the players from the production of the raw materials up to the consumption and interplay with many food-related ones (e.g., packaging, engineering, biotechnology), is recognized as being one of the most important globally, and its status, development, and economic value strongly depend on natural, human, and physical resources (Genier et al. 2009).

In the last decades, growing issues related to food security, animal welfare, environmental events, and climate changes, along with social (labor) conditions of the food production of the entire food value chain, led to the definition of a complex set of requirements to meet the expectations of the consumers and the civil society in terms of quality, healthiness, and safety of products. The increasing complexity of the current global food system at all stages, from the agricultural production to food consumption, demands that the food industry pays main attention to the implementation of ethically sound practices (Coff 2013; Mepham 2013; Mepham 2010). In this framework, food ethics has become a relevant area of interdisciplinary discussion on major current and emerging issues of the food production including, among others, traceability and communication (Gottwald et al. 2010; Coff et al. 2008).

An ethical and responsible food chain implies a two-level approach. The first one is associated with the diverse entities and stakeholders acting in the food value and

P. Pittia (✉)
Faculty of Bioscience and Technology for Food, Agriculture and Environment,
University of Teramo, Teramo, Italy
e-mail: ppittia@unite.it

© Springer International Publishing AG 2018
R. Costa, P. Pittia (eds.), *Food Ethics Education*, Integrating Food Science and
Engineering Knowledge Into the Food Chain 13,
https://doi.org/10.1007/978-3-319-64738-8_6

production chain including enterprises, farms, distributors, services, as well as all the organizations and associations, of public or private nature. Food companies, especially those of large dimensions (e.g., multinationals), in the attempt to communicate their ethical and responsible status to the end users (consumers, other enterprises, etc.) of their products, started to disseminate their management and corporate social responsibility strategies via public documents and other contents made available on their websites.

The second one is related to the individuals that, at various levels and in different sectors, are involved in the activities of the food production, from the development of new knowledge (e.g., scientists and researchers) to the manufacturing, distribution, and consumption as well as their control of the food materials and products. Depending on their role and activity, we could mention here entrepreneurs, food practitioners or professionals, employees, and workers of any food production-related activities, but also teachers and scientists. It has been recognized that the availability of qualified knowledge-/science-based and appropriately motivated employees may play a major role in improving the status of the food and drink industry (Oreopoulou et al. 2015). The complexity of food processing is continuously increasing, and the knowledge of many scientific disciplines (e.g., nutrition, safety, environmental studies) is required by the future food scientists and engineers (Dumoulin 2012).

However, besides technical skills and knowledge, any individual and professional, employee or employer involved in food production, is charged with ethical responsibilities and is expected to apply ethical approaches and skills in their work. It is clearly evident that any practice that configures unethical behavior as a result of ignorance, lack of the required scientific knowledge and technical skills, lack of freedom, lack of approval (Seebauer 2013), mismanagement, and misconduct, may have crucial consequences at various levels and, at large, on our society, with criminal consequences when the minimum standards of quality and safety of foods are not abided.

At the academic level, food study programs aim at developing the main generic, scientific, and technical skills on food science and technology or engineering, while scarce attention is given to teaching activities addressed to set ethical principles and to implement an ethical mind-set of the students and future food professionals.

In a framework where reference rules and guidelines are lacking, an instinctive ethical behavior from any people involved in the food chain is expected, complemented by the duty to comply with the regulations and a vast food law taken as reference to identify and determine any criminal actions (fraud, sophistications, etc.). Ethical behavior, however, is far from the simple respect of legal requirements, and a wider and comprehensive approach that includes also the impact of any professional and working activity on our society is needed.

The lack of references for the ethical behavior of food professionals is partly filled by more or less formal documents generally named as code of ethics (CoE) emanated by entities of various nature including professional societies, corporations, and associations. For their members, the corresponding CoE represents both a valuable tool of professional integrity and a reference of the ethical standards expected in the workplace (Olson 1998) or framework of action to be transferred and instilled to the new members. Moreover, a CoE may contribute to enhance the judgment ability and ethical sensitivity, to support and strengthen individual's moral and an organization's sense of identity.

On the other hand, where the food profession is regulated by law (e.g., Italy, Cyprus, Turkey), the CoE serves, in general, as a basis for taking formal or legal disciplinary actions against violators within the specific framework.

6.2   General Aspects

The term "code of ethics" has several definitions. In the Business Dictionary, it is intended as a "written set of guidelines issued by an organization to its workers and management to help them conduct their actions in accordance with its primary values and ethical standards" (Business Dictionary 2017).

On the other hand, the International Good Practice Guidance (2007) describes it as "principles, values, standards, or rules of behavior that guide the decisions, procedures and systems of an organization in a way that (a) contributes to the welfare of its key stakeholders, and (b) respects the rights of all constituents affected by its operations." In the latter definition, it is highlighted the main impact that the personal ethical behavior and the respect for the principles can have on the specific system where the individual is acting and the overall relationships.

Various could be the aims and areas of applicability, making more complex a general definition and description of a CoE.

In general, a CoE is a reference guide of principles issued by institutions or business entities, of private or public nature, to assist their members, employees, and professionals in performing their activities, or businesses, honestly and with integrity. A code of ethics may also outline the mission and values of any organization or company, the ethical behavioral principles based on its core values and standards, and how its members and any other interested and related individual intended to approach problems within the same entity. The CoE sets out the rules for professional behavior, and any action that does not comply with its contents may determine termination of the position or the dismissal.

For public governmental or nongovernmental organizations, like professional orders, chambers, and boards charged with the management of officially regulated professions, the CoE is a compulsory tool that regulates the practice of the profession itself, by setting the principles and standards of morals and behaviors. It is intended as a document of professional responsibility and, in general, includes guidelines to face difficult issues and decisions that need to be made about what is right or wrong and provides indications of which behavior is considered "ethical" under the specific professional or working circumstances. The moral responsibility of the group to whom it is addressed is usually defined in the "statement of purpose" and included in either the preamble or the declaration of intent (Olson 1998).

In the business and professional sectors within the same enterprise, a CoE could be issued at various levels, and the corresponding documents could regulate ethical aspects at different levels and be addressed to three target groups: business, employees, and professional practice.

Some companies and organizations while having a reference CoE develop and adopt a corporate code of conduct (CoC). Sometimes the two documents (CoE and CoC) are mixed, and the terms "ethical code" and "code of conduct" are wrongly considered as interchangeable terms, while a clear distinction among the two does exist. A CoE is a public document that sets out the values and ethical principles underpinning the code itself and includes a series of principles that affect decision-making of any individual to whom it is addressed. A CoC is, on the contrary, a document addressed to and intended only for employees or members of the company and defines rules and restrictions and regulates their behavior within the specific environment and, thus, is addressed and made available only to them.

Diverse CoE documents have been issued in various disciplines and professional and business areas as well as by various groups of individuals (associations, networks) due to the intrinsic and required specificity related to the principles, values, and morally permissible standards of conduct to which members of a group bind themselves. A CoE could also prioritize the principles either by explicitly weighting the principles or by implicitly ordering them in the document.

In the development and writing of a CoE, some aspects have to be taken into account and in particular:

- Who are the individuals or the group of individuals to whom it is addressed to and which are the ethical priorities
- Which are the main areas of action of the group
- Which unethical decisions have to be prevented and how
- Which type of ethical problems the members of the group may encounter in their work/profession/activity
- Any conflicting principles and how to resolve

Based on these premises, it is evident that a code of ethics for a food professional will contain principles different from those included in the same type of document developed for other professions and professionals (e.g., civil engineer, biologist, chemists).

A general classification of the code of ethics could be based on how the above-listed aspects are organized and prioritized in the text or as a function of the area of interest or activity (e.g., healthcare vs. engineering, science and research vs. education).

Other classifications do exist, and, as an example, Olson (1998) proposed one according to which any CoE could be classified as one the following three main models:

1. Brief model: This class includes CoE with a small amount of information and a short list of statements, without the main structure.

2. Principle model: In this category, included are documents that focus on specific principles that the member/individual has to respect and to apply to the potential relationships the members of the group may encounter. The text is developed on sections where principles (fundamentals), canons, and how they have to be respected (guidelines for the principles and canons) are described.
3. Relationship model: In this case, the contents of the document highlight the relationship between the individual/group and other groups of the society and stakeholders (e.g., public, consumers, clients, etc.). In general, it comprises sections with headings like "Relationship/obligations to …" followed by a list of standards and guidelines.

Another general classification is based on the "nature" of the code, and a possible classification is grouping codes into the following groups (UNESCO 2006):

- Aspirational: documents with a list of the statement of virtues to which practitioners should strive. They are focusing on the fullest realization of human achievement instead on notions of "right" and "wrong."
- Educational: a code that serves to enhance understanding of its provisions with extensive commentary and interpretation.
- Regulatory: a code that includes a set of detailed rules to govern professional conduct and to serve as a basis for adjudicating grievances. Such rules could be enforceable through a system of monitoring and the application of a range of sanctions (Frankel 1989).

The main challenge and characteristics of a CoE are:

- Be applicable to any situation the member/individual may encounter within the framework of activity.
- Be understandable and of easy application.

To set a CoE with these characteristics, the involvement of as many individuals as possible in making and accomplishing it is an essential aspect. It allows the collection of the needs of all, the identification of any critical issue, and the tune and harmonization of the final document within a general and shared consensus.

6.3 Codes of Ethics in the Food Science and Technology Sector

The respect of ethical principles by any food professional involved in the food chain is an intrinsic behavioral principle and part of his/her personal integrity due to the main impact that actions have on health and well-being of all humans and, overall, our society. In these last decades, the importance of ethics, virtues, and values in the food sector has increased due to growing societal issues in relation to innovations, novel foods, food security, safety, as well as sustainability and the environmental impact of the food productions. A series of books on the general topic "food ethics" have been recently published, and this attests the increasing interest to develop a "value-based" approach of the food processing and the need of tools to generate a higher consciousness at the national and global level in the wider scientific and professional area (Thompson 2007; Coff et al. 2008; Gottwald et al. 2010; Clark and Ritson 2013; Dumitras et al. 2015).

Despite their importance, it is out of the scope of this chapter to discuss about the main and relevant topics of the food ethics (e.g., GMOs, food security, malnutrition), while, on the contrary, the attention here is focused on the tools. And inter alia, the code of ethics is emanated and adopted by organizations and institutions to set out the values and the ethical behavior of the individuals to whom their activities are addressed to (food professionals, employees, etc.). In the same framework, but for different purposes, many food companies and private and business entities have developed a code of conduct to translate the ethics code into specific guidelines and to promote adherence to the code of ethics.

6.3.1   Food Enterprises and Business

Food companies can develop a code of conduct (CoC) (in some cases also named as "code of business conduct") as part of their corporate governance. In general, it is a directional document, including a series of rules to regulate the standards of their corporate behavior toward everyone they work with, the communities they get in touch, and the environment on which they have an impact. Thus, CoC contents vary depending on various factors (e.g., company size, geographical business area, sector).

These documents set the guidelines and rules designed to outline specific practices and behaviors that influence employee's actions and that should be followed or restricted under the company in its general activities (e.g., employees' relationship, suppliers' standards, food safety, public health). In general, they apply also to the management, business units, and subsidiaries and, in some cases, also to individuals out of the company with whom the company may collaborate (e.g., consultants, lobbyists, brokers, agents, and other representatives of the company).

The code of conduct documents are addressed, in general, to any individual within the company or connected with it, to regulate the specific activity and, thus, are specific and private.

However, increasing is the number of the corporates and, in particular, the large and multinational ones, whose CoC documents can also contain their core ethical values, principles, and ideals. In this case, it could be publicly disclosed to the wider community of the consumers/end users with the aim to inform them about the corporate ethical approach and improve their trust on the food company.It can also serve as guidelines for making decisions in their business, private activities.

6.3.2 Organizations Which Adopted Code of Ethics for Food Science and Technology Professionals

In the food science and technology sector, several organizations/associations have developed a CoE to regulate the professional behavior of their members.

Based on the information currently available, CoE has been issued by two main types of entities: (a) nonprofit associations and nongovernmental organizations and (b) professional orders/association/chambers/corporations of professionals.

In the former category (a), included are associations and organizations whose individual members enroll either on voluntary basis, depending on the specific interest, or upon selection, or invitation. In this group, the following organizations have issued a CoE for their members:

1. International Union of Food Science and Technology (IUFoST 2017). IUFOST is a global scientific organization in the food science and technology sector supporting activities of various natures to increase the safety and security of the food supply chains at international level. It is a nonprofit institution, and its members are representatives of national food science organizations from different countries worldwide.

2. Institute of Food Technology (IFT 2017). This is a private association of food science and technology professionals born in the USA recognized as the first association of food technologists. It develops networking, learning activities to foster technology development and transfer and to promote and support research, and innovation in food sciences. IFT activities are also aimed to enhance recognition of the food profession and increase understanding and application of the science of food. IFT has developed the Certified Food Scientists (CFS) credential globally recognized as a designation for food science professionals that certifies their applied scientific knowledge and skills.

3. Institute of Food Science and Technology (IFST 2017a). Based in the UK, it is the only professional body concerned with all aspects of food science and technology (individual membership), but, at a larger scale, it acts as an independent qualifying body for food professionals in Europe. It includes members of varied backgrounds including industry, academia, government, research and development, quality assurance, and food law.

4. New Zealand Institute of Food Technology (NZIFST 2017). This organization, founded in 1965, was initially mainly linked to the emerging discipline of food technology at Massey University and its graduates, while today the membership comprises food technologists/engineers and scientists from many disciplines and from diverse educational backgrounds. NZIFST introduced in 1968 a code of ethics for its members with the aim of keeping ethics and excellence in the sector and of safeguarding standards in the profession.

5. ISEKI-Food Association (IFA 2017). This is a nonprofit, international organization born in 2005 as a network of scientists, teachers, professionals, and industry representatives with activities aimed to bridge education, research, and business.

It has recently issued a code of conduct for its members (2017) based on the *Code of Ethics for Food Science and Technology Professionals*, developed under the project Track Fast (Training Requirements and Careers for Knowledge-based Food Science and Technology in Europe, https://www.trackfast.eu) with the scope of keeping high the ethical and professional values of its members.

In the (b) category (professional orders/association/chambers/corporations of professionals), organizations and entities, officially recognized by public authorities and whose existence depends on national rules and laws (for Europe, Directive, 2005/36/EC of the European parliament and of the council of 7 September 2005 on the recognition of professional qualifications), are included. The food professional orders and chambers aim to regulating the profession of their members as food technologists/engineers and/or other food-related professions based on the certified or recognized skills and knowledge as well as to acknowledge and socially to recognize the impact of the work activities of food technologists on public health. The members of the chambers/orders have to respect a code of conduct that imposes an ethical approach in the work activity (Costa et al. 2014).

In this category, only the entities dealing specifically with "food technology/ engineering" professions and activities will be considered and, thus, the food-related ones like (food) nutrition, medicine, and chemistry, will be excluded.

Despite the worldwide presence of academic programs that confer a degree with various titles on food studies (e.g., food science, food science and technology, food engineering, etc.) that allow entering professionally in the corresponding work sector, only few countries had set orders or chambers of food technologists or engineers according to national and international laws.

Currently, in Europe, the following food professional bodies, with a sector-specific CoE, have been identified:

1. Italian Professional Order of Food Technologists (Ordine Nazionale dei Tecnologi Alimentari): public body set under the Ministry of Justice, laws that officially regulate the profession of food technologists according to national and European laws.
2. Association of Food Technologists (Cyprus): this is a regulatory institution in the country, and enrolment (with membership fee) for FS&T professionals is compulsory for consultancy and professional activities.

It is worthy to mention that in some countries (e.g., Turkey or Greece) the food technology profession does not exist per se while it is included as a "sector-specific" ones. Thus, food professionals are members of more wide and general orders/chambers (e.g., engineering or chemistry). In these cases, they have to respect rules and guidelines of code of ethics of more wider and general applicability.

Another case is the Icelandic Food Scientist (*Matvælafræðingar*), regulated according to the EU Directive (recognition under Directive 2005/36/EC) and included in the list of other authorized "health professions." In the Healthcare Practitioners Act No. 34/2012 issued by the Ministry of Welfare to officially regulate the professions in the health sector, a section is dedicated to the rights and obligations of healthcare practitioners with no distinctions among the authorized health professions and including general indications on the duty of respect for professional standards and responsibility.

6.4   Aims and Contents of the Code of Ethics

6.4.1   Aims

The various institutions and organizations that so far have issued and adopted CoE for their members claim different reasons for the existence of this document.

IUFOST, as an internationally recognized organization including, among the adhering member's research, professional institutions and organization from all over the world, in its CoE states about its "responsibility for establishing standards of professional practice for the guidance of its adhering bodies."

Both IFT and IFST, organizations "profession"-oriented, highlight the importance of this document to "assure the integrity, honor, and dignity of the Institute and its members." The Institute of Food Technology has also defined an additional and specific code of ethics for its members (scientists and professionals) that receive the IFT certification as food scientists with additional and more specific guidelines.

On the other hand, IFST recognizes the importance of having a CoE with respect to both other professional bodies, governmental authorities, and departments and employers of food scientists and technologists, for its potential impact on the wider community and public. It is regarded as useful tool to benchmark the ethical standards of their members. This statement reflects the fact that in the UK, despite food science and technology is a well-recognized discipline and qualified academic study programs are included in university departments and other higher education institutes, only in relatively recent years, it has received acceptance as a profession. In this respect, IFST has adopted a code of professional conduct for the food technologists with guidelines to guarantee their social and public responsibility for the wholesomeness of food and integrity toward the profession. IFST is also committed to the development of additional activities in the UK, in collaboration with other institutions, aimed to certify the skills and good knowledge of the food manufacture, of the quality principles, and their application to all sectors of the food industry.

The ISEKI-Food Association has recently issued for its members a code of ethics for the food professionals as a tool for making the food professionals aware of their duties to society, increasing consumers trust in the food supply chain, and improving the social recognition of the profession in any field. Within the philosophy and aims of the association, the code is made available also to be adopted by any professional organization at the international or national level.

Different are the goals of CoE documents adopted by public and officially recognized authorities, bodies, and organizations like the professional orders (like in Italy and Cyprus). For these entities, this is the sole document their members have to refer to and the ethical benchmark to which they have compulsory to comply with. In general, this document includes, besides the main values of the professionals, the main commitments of the individual to the society and to the profession as a member, with a detailed description of the specific duties within the organization, with the other members and toward the customers or end users, similar to a code of conduct. The principles and rules set out in the CoE are binding on all members of the body.

6.4.2   Contents

It is recognized as the main impact of any actor and activity carried out within the food supply chain on the public health and well-being. Thereby, a general ethical approach and a professional integrity are, thus expected by any individual operating in this sector with the main reference to the respect of public laws, the correct control of the food processes to assure quality and safety of the products and the commitment to guarantee food for all (food security). Furthermore, any activity should be aimed to limit the impact of the food production chain on our society and environment and overall to improve the sustainability of the food value chain and to contribute to a responsible innovation.

These concepts have been differently implemented in the existing CoE of the food technology sector depending on the aims of the organization, institute, or public entity. Differences could be evidenced regarding the model used (brief, principle, etc.) as well as on how the contents have been developed and described.

For the scope of this chapter, the date/time of issue of the document was considered as a possible criterion to analyze any change of values and principles of the food sector during the years that may have led, in turn, to changes in the CoE approaches and corresponding guidelines and contents and/or to compare different versions and updates. However, the available references did not allow to obtain all the corresponding information, and thus this analysis is referred only to the current documents available to the public at the time of writing of this chapter.

The IFST (2017b) and the Italian Food Technology professional order (2017) have a Code of Professional Ethics based on a relationship model, and their document includes a series of titles and articles to regulate the professional behavior and relationship with colleagues, customers, and impact on the society.

In the IFST's CoE, the duties are described in 10 guidelines with the following titles: Wholesomeness of food, Relation with the media, Confidentiality of information, Conflicts involving professional ethics, Duties towards subordinates, Scientific issues and food promotion, Responsibilities towards students, Responsibilities to the environment, Members business interests, and Responsibility to the profession and the institute. This model complies with the role of this organization and the official recognition of the profession in the country.

The CoE of the Italian Order of Food Technologists, besides a preamble and a closing part, includes 33 articles categorized under three sessions with the following titles: Relationship with the other professionals, Relationship with the customers, and Relationships with the public and society.

Other organizations, like IUFOST and IFT, have adopted a brief model, and the document includes only general statements that the adhering members of these organizations shall respect.

The compliance of the professional activity of the food technologists to ethical principles of conduct is of main importance when society, environment, and consumers are taken into account. In general, these arguments are all included in all the CoE here considered, with some differences among them in terms of the relative importance of the various aspects and/or how their specific guidelines are detailed.

Among others, three most important aspects dealing with the ethical principles and moral conduct of food technologists and professionals will be here deepened due to their main importance in relation to the contents of this book where this chapter is included.

6.4.2.1   Responsibilities on Food Ethics Concepts

Duties and responsibilities of the activities of food technologist and professionals on various food ethics aspects are developed to a different extent in all the available CoE. In a condensed style, IFT includes the statement "Members of the IFT shall work to ensure the health, safety, and well-being of the public."

The documents developed by the other organizations under analysis in this chapter include a more or less detailed description of the duties and responsibilities on various aspects of food ethics (food safety, food security, valorization of food technologies, food waste, sustainability) as well as the potential impact of the respective actions on consumers, society, and environment. The main aim of these articles is to provide guidelines for the professional activity to guarantee the protection of consumers' health through the assurance of food wholesomeness. The CoE of the Italian Order of Food Technologists states the commitment of professionals to innovation of food products, processes, and technologies thereby encouraging the economic and cultural growth along with the development of a knowledge-based society.

6.4.2.2   Responsibilities as Researchers and Scientists

The ethical behavior of the food technologists as scientist and researchers is taken into account at various levels. All CoE include the duty and commitment of the food technologists as master and leading innovators in the field.[1]

---

[1] See https://ec.europa.eu/programmes/horizon2020/en/h2020-section/responsible-research-innovation.

The available documents do not contain references to the obligation of development responsible research and innovation according to the European and international approaches as well as to "scientific integrity," while, in general, CoE is more focused on issues related to specific activities that may require ethical approaches especially when dissemination and communication of scientific results and knowledge are considered.

The IFT document focuses on ethical rules for scientists as authors of publications and indicates that its members shall "report all scientific research properly and accurately," "acknowledge the work and publications of others properly and accurately," "maintain objectivity when reviewing scientific work, publications or journals," and "not plagiarize the research of others or use the research of others without proper authorization." IUFOST, in its brief Code of Professional Conduct, recalls the "integrity of their professional publications whether through the medium of the spoken, written or printed word, or by radio or television broadcasts or by any other means."

6.4.2.3 Responsibilities as Trainers and Educators

The role of education and training in the skills and competences of the food technologist is well recognized as a key baseline for any work and profession, and the duty of continual professional development is a fundamental element of the CoE adopted by professional orders and organizations of professionals (e.g., IFST, NZIFST).

On the other side, scarce are the references in these documents to duties of the food technologists to education, training, and knowledge transfer. Currently, only the CoE of IFST includes a section focused on the responsibility of the FS&T professionals as teachers or trainers to students. The "responsibility (of the IFST member) to assist their development" is included along with specific sessions dedicated to the responsibilities of academics and of managers during students' industrial training.

6.5 Conclusions and Perspectives

Code of ethics issued for food technologists and professionals by organizations and public entities confirms the importance of the ethical behavior, morality, and integrity in all the activities of the food value chain.

They represent a benchmark tool that may contribute and guide any action and activity made and developed within the work, profession, and everyday life to guarantee wholesomeness, safety, and quality of foods to the consumers as well as to promote the sustainability and innovation of the food value chain.

A revision of the available CoE should be considered as a regular action to adapt and upgrade the documents to the continuously growing issues in the food production and the changing importance of the food sector in the manufacturing sector as well as in our society.

In this regard, specific guidelines on the ethical values and duties of the food technologists and professionals as trainers should be adopted in all food professionals' CoE due to the growing role that education has in the development and innovation of the food sector and of a knowledge-based society.

References

Business Dictionary (2017) WebFinance.inc, Fairfax. www.businessdictionary.com/definition/code-of-ethics.html. Accessed 15 Jan 2017

Clark JP, Ritson C (eds) (2013) Practical ethics for food professionals. Ethics in research, education and the workplace. IFT Press-Wiley-Blackwell, Oxford

Coff C (2013) A semiotic approach to food and ethics in everyday life. J Agric Environ Ethics 26:813–825

Coff C, Barling D, Korthals M, Nielsen T (eds) (2008) Ethical traceability and communicating food. Springer, Dordrecht

Costa R, Smole Mozina S, Pittia P (2014) The regulation of food science and technology professions in Europe. Int J Food Stud 3:125–135

Dumitras DE, Jitea IM, Aerts S (eds) (2015) Know your food. Food ethics and innovation. Wageningen Academic Publishers, Wageningen

Dumoulin E (2012) Changes and perspectives in food studies. Int J Food Stud 1:211–221. http://dx.doi.org/10.7455/ijfs/1.2.2012.a10

Frankel MS (1989) Professional codes: why, how, and with what impact? J Bus Ethics 8:109–115

Genier C, Stamp M, Pfitzer M (2009) Corporate social responsibility for agro-industries development. In: Da Silva C, Baker D, Shepherd A, Jenane C, Miranda-da-Cruz S (eds) Agro-industries for development. CABI, Oxfordshire

Gottwald F-T, Ingensiep HW, Meinhardt M (eds) (2010) Food ethics. Springer, New York

IFA, ISEKI-Food Association (2017) www.iseki-food.net. Accessed 10 Jan 2017

IFST(2017a) Institute of food science and technology. www.ifst.org/. Accessed 10 Jan 2017

IFST (2017b) Professional conduct. http://www.ifst.org/about/professional-ethics. Accessed 10 Jan 2017

IFT, Institute of Food Technology (2017.) www.ift.org. Accessed 10 Jan 2017

Italian Food Technology professional order (2017). Codice deontologico. http://www.tecnologialimentari.it/it/consiglio-nazionale/codice-deontologico/. Accessed 10 Jan 2017

IUFOST, International Union of Food Science and Technology (2017) www.iufost.org. Accessed 10 Jan 2017

Mepham B (2010) The ethical matrix as a tool in policy interventions: the obesity crisis. In: Gottwald F-T, Ingensiep HW, Meinhardt M (eds) Food ethics. Springer, New York, pp 17–30

Mepham B (2013) Ethical principles and the ethical matrix. In: Clark JP, Ritson C (eds) Practical ethics for food professionals: ethics in research, education and the workplace, 1st edn. John Wiley & Sons, Ltd., Hoboken, pp 39–56

NZIFST, New Zealand Institute of Food Technology (2017) www.nzifst.org. Accessed 10 Jan 2017

Olson A (1998) Ethics code collection. http://ethics.iit.edu/ecodes/authoring-code

Oreopoulou V, Giannou V, Lakner Z, Pittia P, Mayor L, Silva CL, Costa R (2015) Career path of food science and technology professionals: entry to the world of work. Trends Food Sci Technol 42(2):183–192

Seebauer EG (2013) Fundamentals of ethics: the use of virtues. In: Clark JP, Ritson C (eds) Practical ethics for food professionals. Ethics in research, education and the workplace. IFT Press-Wiley-Blackwell, Oxford, pp 3–20

Thompson PB (ed) (2007) Food biotechnology in ethical perspective. Springer, New York

UNESCO (2006) Division of ethics of science and technology. Interim analysis of codes of conduct and codes of ethics, September 2006

Chapter 7
Corporate Social Responsibility

Louise Manning

7.1 Introduction

Corporate social responsibility (CSR) is a phrase that has gained wide interest in recent years. CSR as a concept does not sit in isolation. Corporate approaches interface with the social responsibility of individuals, both internal and external to the organization and also of governments too. This chapter will begin by outlining the background surrounding, and the definitions of CSR, and then detail how CSR is adopted in food supply chains at a strategic and operational level. This chapter has been written to provide support to educators who engage learners in the wider subject of food ethics and also as a point of reference for students.

The role of education is to encourage learners to independently think, to analyze and evaluate material and evidence from a range of sources, and to ultimately develop the ability to critically reflect and produce compelling arguments around the subject matter they are engaging with. When considering food ethics, there is rarely only one uniquely correct answer. In fact when deliberating on questions of morality, the lives we live influence our opinions and judgments and, thus, the arguments that we form. Therefore based on individual beliefs and cultural norms, there are varied points of view on any given topic, and all views can be deemed valid if, and only if, they are supported with appropriate objective evidence which is drawn together in a cohesive, persuasive manner.

Evidence whether presented in the literature as an academic paper or a corporate annual report or indeed as a vocal statement is a body of facts or information that supports and gives strength to ideas, assertions, or a point of view. Stating a fact as if that process in itself gives it kudos and value is simply not enough when constructing an academic argument.

L. Manning (✉)
Harper Adams University, Newport, Shropshire, UK
e-mail: lmanning@harper-adams.ac.uk

© Springer International Publishing AG 2018
R. Costa, P. Pittia (eds.), *Food Ethics Education*, Integrating Food Science and Engineering Knowledge Into the Food Chain 13,
https://doi.org/10.1007/978-3-319-64738-8_7

Table 7.1 Characteristics of three kinds of argument

Objective argument	Rhetorical argument	Eristic argument
One truth	Several interpretations	One "party line"
Necessary or probable	Justifiable	Imposed by threats, fear, or power
Aims to convince	Aims to persuade	Aims to compel

Source: Russell and Greenhalgh (2009)

7.2 Rhetoric and Argument

Academic, and as equally corporate, rhetoric, especially in the area of CSR, is often about a persuasive discourse. Rhetoric uses the powers of persuasion to influence others and is an element of overall communication with both *analytical* approaches (focusing on logic) and *dialectic* (debating a point or issue) elements. Persuasion is said to include three elements: *logos*, the argument; *pathos*, the appeal to values and beliefs; and *ethos*, credibility (Van de Ven and Schomaker 2002). Understanding firstly the definitions and also the interplay between them in effective, or conversely ineffective, communication underpins any engagement with CSR literature. Russell and Greenhalgh (2009) drew together the characteristics of three types of argument that are all seen in the conceptual space of CSR (Table 7.1). It is important when learners go on to read corporate literature and policy documents that they clearly understand the difference between these three types of argument.

7.2.1 Objective Argument

Objective argument suggests there is only one solution, one truth, and is aiming to convince the target audience(s) with the piece of communication that this is the case.

7.2.2 Rhetorical Argument

Rhetorical argument can be interpreted in several ways, e.g., the phrase "tough on crime; tough on the causes of crime." It appeals to two audiences with different viewpoints at the same time because the start of the phrase is conservative in nature and the end has a liberal focus. This type of persuasive rhetoric implies that the communicator is a friend of all and understands and engages with all points of view. Some would call this *strategic ambiguity*, i.e., those instances where individuals either as themselves or on behalf of corporations purposefully use ambiguity in order to accomplish their goals (adapted from Eisenberg 1984). This approach can suggest, as Eisenberg indicates, a form of unified diversity, i.e., through ambiguity all multiple viewpoints are met, and approaches can be flexible as long as the goal is reached. This notion purports that organizational mission, goals, and plans can be achieved, even if there is an implicit vagueness in the communication, as individuals will then align, so there is one collective voice. Sim and Fernando (2010) argue that strategic ambiguity can be used for four reasons:

1. Promoting unified diversity
2. Preserving privileged positions
3. Promoting deniability
4. Facilitating organizational change

Strategic ambiguity is a tactic that is utilized in the food industry, often for good motives. However, Sim and Fernando (2010) assert that the influence of strategic ambiguity on both organizational strategy making and communication may lead to a challenge of unethical executive behavior and action, and this is sometimes determined to be the case. In the United Kingdom (UK), the Advertising Standards Authority (ASA) is asked to rule on communications in the press as to whether they are or they are not, considered as misleading. In 2013 Tesco Stores Ltd. issued an advertisement "What burgers have taught us" in all the major UK newspapers in the wake of the European horsemeat scandal. It contained the sentence:

It's about the whole industry. (ASA n.d.)

Box 7.1 Tesco Advertisement Following the 2013 Horsemeat Scandal
What burgers have taught us
The problem we've had with some of our meat lately
is about more than burgers and bolognese.
It's about some of the ways we get meat to your dinner table.
It's about the whole food industry.
And it has made us realize, we really do need to make it better.
We've been working on it, but we need to
keep going, go further, move quicker.
We know that our supply chain is too complicated.
So we're making it simpler.
We know that the more we work with British farmers the better.
We've already made sure that all our beef is from the UK and Ireland.
And now we're moving onto fresh chickens.
By July, they'll all be from UK farms too. No exceptions.
For farmers to do what they do best,
they need to know they've got our support.
We know this because of the work we've been doing
with our dairy farmers to make sure they always get paid
above the market price.
We know that no matter what you spend,
everyone deserves to eat well.
We know that all this will only work if we are
open about what we do.
And if you're not happy, tell us.
Seriously.
This is it.
We are changing.
Source: The Grocer 2013

Two complaints were received: one from an independent butcher stating that this claim was misleading because "they believed it implied there were issues with meat standards across the whole food industry" (ASA n.d.). The ASA's assessment was that they noted "Tesco's assertion that the ad would have been interpreted as a reference solely to their own contamination issues. We considered that despite the use of words such as "we" and "our" in the preceding sentences, the ad made a definitive statement, "It's about the whole food industry". We considered that the omission of "we" or "our" from that sentence made it stand out from the surrounding text and informed readers' understanding of the rest of the ad. Therefore, we concluded consumers would understand the ad referenced all food retailers and suppliers, rather than Tesco alone" (ASA n.d.). The text from the whole advertisement is in Box 7.1.

Considering this statement in light of the four reasons for strategic ambiguity identified above, it could be argued that the statement was designed to facilitate and communicate: promoting unified diversity through the phrase "We know that no matter what you spend everyone deserves to eat well" and facilitating organizational change, e.g., "… we need to keep going, go further, move quicker.… Seriously. This is it we are changing." The vocabulary used spoke to a range of stakeholders, consumers, farmers, shareholders, etc. Sim and Fernando (2010) cite Wexler (2009) suggest that in crisis situations, there are a number of roles that strategic ambiguity addresses including:

- Sealing or buffering part of the organization from closer scrutiny
- Seeking deniability rooted in earlier ambiguous communications
- Addressing issues that are controversial, divisive, and likely to lead to conflict or escalating demands for action
- Buying time when pressed for a decision or specific information
- Fostering organizational change requiring buy-in from stakeholders

The latter clearly being a role of the news advertisement in the case described.

Thus, strategic ambiguity allows multiple perspectives and objectives to coexist (Leitch and Davenport 2007). The value of ambiguous communication is clear in a situation where what is believed to be true is fluid or subject to levels of uncertainty, or different groups may have a different *definition* of what they believe to be true. For example, the word "sustainable" can be acceptable to many stakeholders when there are multiple views as to what it actually means (Table 7.2).

The Brundtland Commission of the United Nations definition of sustainable development (WCED 1987) is such an example of strategic ambiguity, namely, "economic and social development that meets the needs of the current generation without undermining the ability of future generations to meet their own needs." Maynard and Nault (2005) determined that the term "sustainability" is loaded with vagueness and ripe with contradictions. Pretty (1994) concurs suggesting that any attempt precisely to define sustainability is flawed. Franks (2014) argues that it is the very fact that there is a lack of agreement over the precise definition of sustainability that has allowed academics from many disciplines to contribute to its study. Ultimately strategically ambiguous messages are designed to engender diverse interpretations between varied audience segments (Smith et al. 2006). Further they

Table 7.2 Definitions of sustainability and sustainable development

Source	Definition
WCED (1987)	Sustainable development: economic and social development that meets the needs of the current generation without undermining the ability of future generations to meet their own needs
Lynam and Herdt (1989)	Sustainability: the capacity of system to maintain output at a level approximately equal to or greater than its historical average, with the approximation determined by the historical level of variability. Sustainability is first defined at the highest system level and then proceeds downward; and as a corollary, the sustainability of a system is not necessarily dependent on the sustainability of all its subsystems
Pretty (1994)	Sustainability represents neither a fixed set of practices, or technologies, nor a model to describe or impose on the world. The question of defining what we are trying to achieve is part of the problem, as each individual has different objectives

propose that these different selective perceptions can then translate into a relatively uniform positive corporate image.

7.2.3 Eristic Argument

Eristic argument is refuting another's point of view in order to weaken their opposition. It is an argument based on negativity, differential power, and the use of language to weaken or marginalize others as individuals and with respect to their own argument. An example of this is some of the discourse around obesity implying that individuals are culpable, somehow wholly to blame through greed and sloth, for their own situation. This eristic argument can be used to circumvent suggestions that corporations have a specific moral responsibility and their corporate actions are not without reproach. This argument is at odds with an assertion as to there being a bankrupt ethical underpinning of say a marketing strategy that promotes high-sugar, high-fat products to social groups where it could be argued that their personal social responsibility is in some way compromised, i.e., they are socioeconomically vulnerable such as those individuals with binge eating disorders.

A further example would be a dialogue put forward that differentiates between different stakeholder goals, e.g., with food labeling. An eristic argument toward consumers who individually or collectively may look for product information that is not required by law to be on the food label is that we, the multinational corporation (MNC), do not believe you, the purchaser, need that information. Thus the information requested will not be released until we, the MNC, are legally obliged to do so through an amendment in current legislation. It could be argued that the information requested is of a particular moral significance for those interested purchasers, e.g., whether a food contains genetically modified (GM) materials, is produced from cloned animals, was part of or a whole carcass that has been religiously slaughtered when the consumer does not want to consume meat from religiously slaughtered animals, etc. The eristic argument is about "winning at all costs."

Therefore, eristic activity is associated with compelling or imposing through power what the MNC believe are their goals and objectives. This could be either by direct argument or by a counterargument that may be adopted that the organization or industry has moved so far in a specific direction that it can't turn back. Consumers have the ultimate choice in that they can decide whether or not to buy the product based on their perspective on how the MNC is behaving toward them.

7.2.4   Types of Dialogue

It is important in the study of CSR that students explore, facilitated by their teachers, the different types of primary dialogue that forms the outputs from governments, nongovernmental organizations (NGOs), MNCs, and others. Fairclough and Fairclough (2012) propose that different types of dialogue have their specific collective goals, which between groups can be mutually inclusive or exclusive. However, within the boundaries of these goals for one group, it can be a heterogeneous rather than homologous environment. For example, the group is concerned with the environmental impact of producing a food in a certain way, but not animal welfare, or the group is concerned primarily with the unit cost of a food and not say animal welfare or environmental criteria. Therefore, there is also a range of individual and collective participant goals that influence those participants' (stakeholders) actions in a given situation and with regard to a specific product. Although this argument has its roots in political dialogue, the same analogy with respect to other types of dialogue can be used in discussions of CSR too.

Fairclough and Fairclough (2012), building on the work of Walton (2007), identify a different type of primary dialogue. The concept has been extended to CSR, namely:

- *Deliberation*: interaction at the public level among citizens on how to solve a problem of common concern
- *Eristic*: quarrels acted out among participants who hold a different opinion and where power can be exerted by one participant over others
- *Inquiry (scientific)*: reasoning based solely on known evidence
- *Inquiry (philosophical)*: reasoning that can have evidential elements but can also bring in moral or ethical elements
- *Negotiation:* interaction between two or more groups in order to reach a solution, e.g., work conditions negotiations and supply chain agreements
- *Persuasion:* advocating a range of points of view where one group may attack or criticize others (Fairclough and Fairclough 2012; citing Walton 2007)

The different types of dialogue can be used within CSR communications in either single or multiple mechanisms. Crapanzano (1990) differentiates between such primary and *shadow* dialogue stating that:

> Simply put, for the moment, a **shadow** dialogue is one a speaker has—silently, for the most part—with others, embodied say in his colleagues, who are not present at the **primary** dialogue....

This highlights the external dialogue associated with CSR, i.e., the primary dialogue encapsulated in annual reports and corporate literature and the secondary or shadow dialogue of CSR, a silent, perhaps invisible, internal dialogue that may not represent itself in the visible culture of the organization but can still influence corporate strategy. Culture is the emergent history and traditions that apply meaning to the underlying values and beliefs held by the members of formal and informal social groupings such as an MNC (Buchann and Huczynski 2004 cited by Griffith et al. 2010). Cultures are constantly being interpreted and reinterpreted, produced and reproduced in social relations, suggesting that definitions of culture have moved away from specifying the content or properties of culture (e.g., artifacts, values, basic assumptions), toward a view of culture as an emergent process of reality creation including processes such as collective social identity and the development of shared experiences, memories, and thus personal meanings (Iivari and Abrahamsson 2002). This can lead to a differentiation of subcultures within an MNC and also a differentiation between internal and external dialogue and discourse.

Khatib (1996) defines a subculture as a culture that is separate from the dominant or overarching corporate culture and exists in a department, work group, or geographical location, and its identity includes the core values of the dominant culture plus additional values unique to its members. Subcultures can also be identified through functionalism, e.g., their transactional focus, thus operations, engineering, or technical functions (Manning 2015a). A number of elements impact on organizational culture including:

- *The paradigm*: the mission and values of the organization
- *Organizational structures*: reporting mechanisms and responsibilities
- *Control systems*: the management system which defines the procedures and protocols in place and the internal mechanisms for monitoring performance
- *Power structures*: the decision-making process and identifying where the responsibilities for decision-making exist in practice
- *Symbols*: organizational identity including logos and symbols of power within the organization
- *Rituals and routines*: meeting and reporting structure
- *Stories and myths*: internal business actions and deeds (Johnson 1988)

The existence of, and interaction between, subcultures can cause not only a primary dialogue but also associated multiple shadow dialogues with regard to CSR. It is important to recognize this when considering CSR activities and interfaces not only within MNCs but also in small- and medium-sized businesses.

The literature considers this area of CSR primary and shadow dialogue further. Greenwash is the "selective disclosure of positive information about a company's environmental or social performance, without full disclosure of negative information on these dimensions, so as to create an overly positive corporate image" (Lyon and Maxwell 2011). Delmas and Burbano (2011) propose that more and more firms are engaging in the art of *greenwashing*, through their publications and discourse misleading consumers, whether intentionally or unintentionally, about the organization's environmental performance or the environmental benefits of a product or service. Stigliz (2006) asserts that corporations are adept at image manipulation, speaking strategically in favor of social responsibility even while operationally they continue to evade it. Therefore, he argues that self-regulated CSR is not a strong enough mechanism on its own to drive business behavior. Du et al. (2010) concurred determining that CSR efforts are driven not only by ideological thinking but also by the value that corporations can potentially reap from their CSR endeavors.

7.3   Differentiating Business Morality and Business Ethics

Horner (2003) differentiated between morality, ethics, and law. The definitions used in this work have been adapted in this chapter to a consideration of CSR:

- *Personal and business morality* refers to a set of deeply held, widely shared, and relatively stable beliefs or values within a community or subcommunity. These can relate to geography, social status, religion, culture, etc., e.g., the contentment or otherwise to eat certain animals.
- *Personal and business ethics* refers to values and the justification for right and good actions characterized by virtue, duty, or utilitarian and consequentialist factors. Personal and business ethics not only reflects characteristics of appropriate action but also their application in a given situation.
- *Law* is comprised of "concrete duties established by governments that are necessary for maintaining social order and resolving disputes, as well as for distributing social resources according to what people need or deserve" (Horner 2003: 263).
- *Value* is meeting specific product or service criteria either explicitly or implicitly defined by an individual or group of stakeholders. These criteria when being applied to food can be as widespread as satiety, portion size, convenience, repeatability, cost, and so forth.

Joyner and Payne (2002) considered a historical perspective on the emerging meanings of CSR, ethical or moral considerations, and values (Table 7.3). Comparing six definitions from 1938 to 1984, they determined how corporate culture drives, or conversely does not drive, the ethical behavior of individuals. The word "ethics" is derived from the Greek word "ethos" meaning conduct; customs or character and ethics can be described as the application of morals to human activity (Manning et al. 2006). Food business ethics could be considered to include the following:

- Criminal behavior and the need to be within the law encompassing areas such as food crime, food fraud, and food adulteration
- Fairness and trust especially where these exceed the requirements of the law in terms of specific supply chain standards and practice
- Human values and personal behavior
- Behavior in business
- Behavior in government and policymaking (Manning et al. 2006)

Therefore developing understanding of the concept of ethics in the teaching of, or learning about, food policy is about the application of various moral theories to

Table 7.3 CSR, business ethics, and values: a historical perspective

Source	CSR	Ethical/moral considerations	Values/other
Barnard (1938)	Analyze economic, legal, moral, social, and physical aspects of environment	Morals are active result of accumulated influences on persons evident in actions	Responsibility: power of private code of morals to control individual conduct
Simon (1945)	Organizations must be responsible to community values	Ethical propositions assert "oughts," rather than facts	Firm survival involves adapting objectives to values of customers
Drucker (1954)	Management must consider impact of every business policy upon society	Morality must be principle of action exhibited through tangible behavior	First responsibility to society is to make a profit
Selznick (1957)	Enduring enterprise will contribute to maintenance of community stability	Definition of mission includes wider moral objectives	Leadership requires defense of critical values
Andrews (1987)	Firm should have explicit strategy for support of community institutions	Defining firm only in financial terms leads to subordination of ethical concerns	Ethical behavior is product of values
Freeman (1984)	Business must satisfy multiple stakeholders	Concern for ethics necessary but not sufficient to decide "what we stand for"	Enterprise strategy

Adapted from Joyner and Payne (2002)

the analysis of practical problems. These practical issues in the food industry are myriad, from animal welfare, worker welfare, environmental concerns, resource management, consumer health and well-being, supply chain interactions and business practice, and so forth. Ethics surround an individual's or a communities' ability to have free choice or autonomy, i.e., the capacity to choose freely and be able to direct one's life (Seedhouse 1988). This suggests the human capacity for self-determination or self-governance or to pursue a course of action. Governmental policymaking is often criticized for being paternalistic in nature, i.e., for usurping individual responsibility and interfering in individual decision-making, expressly for the purpose of promoting health and well-being (Buchanan 2008).

Policies promoting choice editing *for* the public, whether food industry generated or at government food policy level, will impact on consumer autonomy (Manning 2015b). This type of choice editing can include retailers only selling free-range eggs, fairly traded bananas, or conversely only selling what would be deemed low ethical value products – intensively produced farm produce or energy-dense and low-nutrient value food choices. The consumer can only purchase what is available, and there are factors of influence that impact on the situation that you can "only buy what is on the shelf," especially in a largely urban population, e.g., socio-economic status, planning regulations, and physical access to food. As previously outlined in this chapter, there is a view by some that individuals are somehow compromised or vulnerable as a result of such corporate activity or ethos.

The concept of *food deserts* explores this element of choice editing for an individual or a community as a result of corporate decision-making, planning decisions in urban areas, low incomes, poor education, or high levels of unemployment forcing specific decisions on what is placed in the food basket. Celnik et al. (2012) from their work in Scotland argue that fast-food outlets are located preferentially in areas of higher deprivation where obesity is more frequent, people are under greater social and economic pressure and are influenced by time stress, and often education standards can be low. This makes negotiation for better access as a community to healthy food options difficult. Choice editing suggests that "others" are better placed than the target group themselves in determining what they need. An individual's approach to risk taking is influenced by a number of factors including the severity of the risk that is being incurred, how that risk is measured against other risks that are situational to their lifestyle, and whether the impact of that risk taking behavior is close or distant in terms of potential fulfillment of the consequences. This again suggests the notion that there are multiple realities within a given group of consumers that influence their decision-making and their interaction with MNCs and their CSR policies. It can be argued that the lack of awareness of specific risks, or just a general acceptance among the socioeconomically disadvantaged of a more high-risk lifestyle, means that they are less accessible to market or policy instruments designed to change food consumption behavior, as can be seen with the higher incident rates of obesity among this specific social group (Manning 2015b).

Craig and Amernic (2004) in their research on Enron concluded that in a reverence to the "market" and a "win-at-all-costs" style of capitalism, the *micro*-discourse suggests that several of the corporate leaders implicated in the collapse of Enron were "deceitful, deceptive, egocentric, arrogant, hubristic and harbored delusional complexes." If society believes that directors of corporations are moral actors and that corporate relationships raise moral as well as economic/financial issues, then a new (or reloaded) discourse is required (Johnson 2003). CSR has edged into this space. Putrevu et al. (2012) consider the issue of corporate social irresponsibility (CSI) although they argue this is nothing new citing Armstrong (1977) and Pearce et al. (2008) who described a management tendency to overlook irresponsible and corrupt practices, but this culture was weaker when leadership was more inclusive and managers were asked to consider the needs of multiple stakeholders.

7.4 Corporate Social Responsibility (CSR)

7.4.1 CSR Definition

CSR was defined by the European Commission as a concept defining how companies integrate social and environmental concerns in their business operations and how they interact with stakeholders on a voluntary basis. However this was updated in 2011 to a simpler definition *the responsibility of enterprises for their impact on society* (EC 2011). Alternatively, CSR is described as the commitment of a business to contribute sustainable economic development, working with employees, their families, the local community, and society at large to improve their quality of life (WBCSD 2003). Other definitions are myriad leading to CSR being seen as a wide and loose concept (Font et al. 2012). CSR is built upon a framework of legislative compliance in conjunction with additional activities and interactions that add value to the organization or the goods and services it provides and benefits either single or multiple groups of stakeholders. The development of CSR standards and benchmarks enables a view to be taken both internally and externally as to the degree of adoption (development, planning, and implementation) of CSR by an organization and its impact (delivery in terms of quantifiable outputs). CSR strategy should be driven not just by ideological thinking; more importantly corporations through philanthropy can also be a powerful and positive force for social change as long as philanthropic goals are aligned with the other corporate goals of the organization (Du et al. 2010). However, too often CSR is considered as a panacea that will solve many global social problems (Van Marrewijk 2003). Scherer and Palazzo (2011) describe five emerging themes of CSR:

1. The emerging global institutional context for CSR: from national to global governance
2. Seeing CSR as a form of self-regulation, i.e., from hard to soft law
3. The expanding scope of responsibility: from liability to social connectedness
4. Evolving corporate legitimacy from cognitive and pragmatic to moral elements
5. CSR translating liberal governmental democracy to deliberative corporate democracy

Garriga and Melé (2004) differentiate between instrumental theories, political theories, integrative theories, and ethical theories of CSR in the following way:

- *Instrumental CSR theories* – focusing on achieving economic objectives through social activity
- *Political CSR theories* – focusing on responsible use of business power in the political arena
- *Integrative CSR theories* – focusing on the integration of social demands, e.g., public responsibility and stakeholder management
- *Ethical CSR theories* – focusing on the right thing to do to achieve a good society

They conclude that these theories revolve around four key aspects of CSR, namely, meeting objectives that produce long-term profits, using business power in a responsible way, integrating social demands, and contributing to good society by doing what is ethically correct. Scherer and Palazzo (2011) distinguish between two of these elements, and the essential characteristics of the instrumental and political approach to CSR have been synthesized (Table 7.4). In the table, the mode of corporate engagement has been moved from legitimacy in the original work of Scherer and Palazzo (2011) to responsibility. The rational for this is that it more clearly

Table 7.4 Characteristics of the instrumental and political approach to CSR

Theme	Instrumental CSR	Political CSR
Governance model		
Main political actor	State	State, civic society, and corporations
Locus of governance	National governance	Global and multilevel governance
Mode of governance	Hierarchy	Heterarchy
Role of economic rationality	Dominance of economic rationality	Domestication of economic rationality
Separation of political and economic spheres	High	Low
Role of law		
Mode of regulation	Governmental regulation	Self-regulation
Dominant rules	Formal rules and "hard" law	Informal rules and "soft" law
Level of obligation	High (enforcement)	Low (voluntary action)
Precision of rules	High	Low
Delegation to third parties	Seldom	Often
Responsibility		
Direction	Retrospective (guilt)	Prospective (solution)
Reason for critique	Direct action	Social connectedness
Sphere of influence	Narrow/local	Broad/global
Mode of corporate engagement	Reactive (as a response to pressure)	Proactive (as a strategic business decision)
Legitimacy		
Pragmatic	High (capitalist approach delivers to the public good via an economic logic)	Medium-low (capitalist approach under pressure)
Cognitive	High (coherent, singular set of assumed business values)	Medium-low (pluralization of often conflicting values according to stakeholder)
Moral	Low moral legitimacy	High-low moral legitimacy (depending on level of discursive engagement)
Democracy		
Model of democracy	Liberal democracy	Deliberative democracy
Concept of politics	Power politics	Discursive politics
Democratic control and legitimacy of corporations	Derived from political system (corporations depoliticized)	Corporate activities subject to democratic control
Mode of corporate governance	Shareholder orientated	Democratic corporate governance

Adapted from Scherer and Palazzo (2011)

describes the corporate view on engagement being either reactive as a response to pressure or proactive with regard to doing what is right, i.e., engaging with CSR because it is the good thing to do rather than because the stakeholders have pressurized the organization to do so.

Proactive vs. reactive CSR is a theme drawn out by Torugsa et al. (2012) who identified an association between proactive CSR and financial performance where there was evidence within the organization of specific capacities in shared vision, stakeholder management, and strategic proactivity. Scherer and Palazzo (2011) conclude that corporations are both economic and political actors. Indeed it could be argued that MNCs often with an economic power larger than some nation states can use this to enact political power. Further in some parts of the world, MNCs operate in countries that could be described as repressive states where as a result the corporate may have one modus operandi, compared to other countries that could be described as "failed states" or "weak states" either due to the reduced economic or political power of the national governments where the modus operandi could be different again.

In this situation, where does the overarching MNC's CSR sit? Is it acceptable for the heterogeneous set of corporate stakeholders that there is a differentiated balance of instrumental to political elements of corporate CSR depending on the specific region of operation? For example, can an MNC have a differentiated set of standards on worker welfare across its global production or supply base? Quite obviously yes, it is a repeated situation across the world. If the MNC actually derives competitive economic advantage from operating in one country where labor costs and expectations as to worker conditions are low in order to export goods to a country where worker conditions are much better, does this situation of economic power override the customer expectation that the MNC's brand is underpinned by a set of core values? These are the kinds of questions that are important for the students to engage with. As was outlined at the beginning of this chapter, there is no uniquely correct answer.

Strategic ambiguity around the exact definition of CSR, its scope, and the interplay between the four elements of CSR leads to a less focused narrative. The narrative surrounding what it is to have sustainable development coexists with "a number of concurrent and oppositional viewpoints" (McGreogor 2004:593). Further he argues that the:

> power of sustainable development is self-sustained through the normalization of particular languages and modes of expression. Alternative ideas and discourses are inhibited by a lack of language and familiarity and consequently disempowered and relegated to subordinate positions within discussions.

This assertion suggests that an eristic narrative sits at the core of the ideas contained within the concepts of sustainable development, CSR, and corporate social performance (CSP). Shareholder value, CSR, and ethics-related concepts are not mutually exclusive; instead they are mutually reinforcing, i.e., shareholder value is essential for CSR, and developing CSR leads to shareholder value (Fassin et al. 2011). According to the scope of the organization's designated social responsibility system, e.g., inclusive or exclusive of nutrition, animal welfare, environmental impact, sustainability (people, planet, profit), etc., the CSR strategy must define the organization's visions, leadership and governance, organizational structure, as well as the associated operational management systems. The strategy should also propose measurable organizational objectives and targets in order to demonstrate to stakeholders actual performance against these objectives and thus the degree of continuous improvement.

7.4.2 CSP: Measuring CSR

Measuring CSR remains a challenge when it is often a social rather than a physical construct (Font et al. 2012). Wood (2010) defined CSP as being a set of descriptive categorizations of business activity, which focus on the impacts and outcomes for society, stakeholders, and also the organization itself. CSP is the measurement of organizational outcomes in the environmental, social, and governance (ESG) domains with respect to multiple stakeholders (Orlitzky et al. 2015). Furthermore the authors outline that CSP drivers operate at organizational, sector, and national levels all within different timescales of influence. Wood (2010) differentiated three aspects of this CSP model:

- *Principles – public responsibility* outcomes, *managerial discretion* in terms of managers and employees being moral actors for the organization, and *legitimacy*, i.e., the way that business uses its power within the confines of societal values
- *Process – issues and public affairs management*, i.e., the processes and dialogue that enable an organization to identify, analyze, and act on the issues that may affect them; *stakeholder management*, i.e., the processes and dialogue that informs active and constructive engagement in stakeholder relations; and *environmental scanning*, collating the information needed to understand the social, political, legal, and ethical environments in which the organization operates
- *Outcomes and impacts of social performance* – i.e., the effect of the organizations' activities and actions on people and organizations, natural and physical environments, and social systems and institutions

Orlitzky et al. (2015) differentiate between six elements of CSP that can be aggregated together to deliver an overall figure for corporate CSP. These elements are customer-orientated CSP, local community-orientated CSP, shareholder-orientated CSP, supplier-orientated CSP, environment-associated CSP, and employee-associated CSP (Table 7.5). They conclude that organizational level factors influence CSP orientated toward local communities, the natural environment, and employees, whereas shareholder dimensions are influenced more by national factors.

CSR is not always associated in an organization to financial performance as the two are not always strategically linked, and there is often a time lag to actually realizing the financial benefit from a specific CSR activity (Joyner and Payne 2002). The balance of evidence is that there is a link between CSR and financial performance, but it is difficult to compare across literature sources because there is a lack of uniformity in measurement, and also the trend may well vary between good economic and poor economic climates (Ducassy 2013). Ducassy (2013), Mahon and Wartick (2012), and others propose a link between corporate reputation (CR) and CSP. However, the latter argue that there is a challenge using a single measure for CSP and CR when the weighting of that measure relates to multiple stakeholder

Table 7.5 Elements of CSP and their characteristics

CSP element	Description	Main components
Customer-focused CSP	This theme provides an overview of the organization's commitment toward maintaining a high quality of products and services and high levels of customer satisfaction and adhering to ethical marketing practices	Quality of management systems Customer satisfaction Competitive practices Marketing practices
Community-focused CSP	Refers primarily to the residents of local communities in which an organization operates. It may also refer to the larger areas, such as a region or nation, to the extent that society in such larger areas is affected by an organization's operations. Examines to what extent the organization takes into account the needs, interests, and rights of communities affected by its operations or planned operations. Pays specific attention to the ways the company seeks to mitigate its negative impact on communities and enhance its positive impact	Stakeholder consultation Processes Contribution to the development of local communities Philanthropic activities Lobbying activities Involvement in nondemocratic countries
Employee-focused CSP	Provides an overview of the organization's commitment toward social issues related to employees, their health and safety, diversity, and employee involvement. For example, whether policies and management systems address core ILO conventions (forced and child labor, freedom of association, right to organize, discrimination)	Working conditions Terms of employment Working environment Industrial relations Employee involvement/participation
Environment-focused CSP	Evaluates the organization's commitment toward the establishment of sound and appropriate environmental management systems (EMS), increasing efficiency in the use of resources and energy, and avoidance of harm to the environment. In assessing each company's environmental record, consideration is given to specific elements that can be categorized under the following headings: Environmental management and reporting systems Organizational record of compliance with applicable environmental laws and regulations Methods of use/extraction of natural resources Level of emissions of hazardous or toxic substances Level of emissions of substances that increase the threat of climate change Organization's impact on natural ecosystems Measures to reduce the environmental impact of operations Ecological impact of the company's products	Resource consumption Air emissions Water and soil releases Waste generation Product impact

Table 7.5 (continued)

CSP element	Description	Main components
Shareholder-focused CSP	Primarily assesses the organization of the board of directors and examines issues such as the independency of directors and the existence and composition of board-specific committees as well as other aspects of good corporate governance, such as transparency, stock ownership structure, voting rights, and compensation paid to senior executives	Independence of directors Audit committee Compensation and remuneration schemes Voting rights Antitakeover devices
Supplier-focused CSP	Refers to the employees of the organization's contractors. It provides an overview of the organization's commitment toward worldwide fair labor standards and freedom of association. The evaluation process examines whether the company implemented a code of conduct that addresses human and labor rights issues relevant to its operations in countries with poor human rights records and whether it implemented the mechanisms to ensure compliance with this code. Controversies include, for example, an organization's complicity in human rights violations, when it is involved directly or through its major suppliers in the use of child, forced, or sweatshop labor	Outsourcing policy Code of conduct for contractors Monitoring of subcontractors and company suppliers Involvement in labor rights violations of firm contractors

Adapted from Orlitzky et al. (2015) citing SiRi Research Framework (2006), SiRi internal documents

perceptions, and these themselves often demonstrate a plurality of views. Joyner and Payne (2002) assert that for an organization to be viable, then CSP and effective financial management need to be integrated as in the triple bottom-line approach (people, planet, profit). Wood and Jones (1995) identify this as the corporate social performance-financial planning (CSP-FP) interaction and suggest four specific elements:

1. Stakeholders are the source of expectations about what constitutes desirable and undesirable business performance. Internal and external stakeholders will influence directly the mission statement of the organization and then the inter-relating aims and objectives that are developed.
2. Stakeholders, especially internal stakeholders, experience the effect of corporate behavior, i.e., they are the recipients of corporate strategy, actions, and output.
3. Stakeholders evaluate how well the organization has met their expectations and/or how organizational strategy has affected the groups and organizations in the external business environment. These evaluations can be quantitative (i.e., measureable), qualitative (relate to perceptions and beliefs), or a combination of both.
4. Stakeholders act upon their particular interests, expectations, experiences, and/or evaluations (Wood 2010). This is especially true of consumers in the food supply chain (Manning 2013).

Therefore it is important to measure CSP in order to get an understanding of the depth of adoption of CSR by the organization.

7.5 The Interaction Between CSR, CSP, and Corporate Financial Performance (CFP)

CSR/CSP can be measured in many ways in terms of *social self-disclosure* such as through voluntary and/or mandatory reporting and *social performance measurement* by the use of a third-party ranking systems or league tables (Font et al. 2012). Third-party ranking systems that are used to demonstrate CSP include the:

- *Global Reporting Initiative* (GRI) launched in 1997 with the goal of enhancing the rigor, quality, and utility of sustainability reporting (Singh et al. 2009)
- *The Institute of Chemical Engineers Sustainability Metrics* published in 2002 (Singh et al. 2009)
- *ISO series of third-party certification standards* and SA 8000
- *Sustainability stock indexes* such as Domini 400 Social Index, Dow Jones Sustainability World Index (DJSI World), Dow Jones Stoxx Sustainability Index (DJSI Stoxx) (see Ziegler and Schröder 2010), and FTSE4GOOD Index

Alternatively, a distinction can be made between hard (normative, i.e., relating to objective measures) and soft (descriptive, i.e., management claims) disclosure (Chen and Delmas 2011; Moroney et al. 2011; Carroll and Shabana 2010).

The hard approach to CSR would suggest that sustainability indicators must be *specific* (outcome bound), *quantitative* (measureable), *usable* (of practical value), *available* (data easily collated), *cost-effective* (not expensive to collect), and *sensitive* (demonstrate changes in circumstances) as identified by Bell and Morse (2003). This does not preclude the use of qualitative indicators, but by their nature, qualitative indications do not readily drive business performance and continuous improvement. Bourlakis et al. (2014) differentiate between four further categories of supply chain sustainability indicators (efficiency, flexibility, responsiveness, and product quality). The literature suggests that both hard and soft approaches to CSR can be adopted within the same CSR strategy.

Sustainability stock indexes can be considered as an indicator of CSP although there are questions about the reliability of this metric (Ziegler and Schröder 2010). Van Stekelenburg et al. (2015) studied the relationship between CFP and CSP (utilizing the DJSI as a measure) concluding that the market rewards organizations with high CSP in terms of stock returns, although the wider literature gives a mixed view. CSP reporting has its own elements of trade-off between positive and negative indicators of performance for the same organization (Chen and Delmas 2011). Therefore, there is often a qualitative assessment or ranking of different CSR issues within a single, or aggregated, CSP metric. There is an implicit desire to find a single measure so that stakeholders can empirically assess business performance and how it changes over time, i.e., a form of corporate profiling (Mahon and Wartick 2011).

If this single performance indicator was developed, it could then be used quantitatively to rank MNCs in terms of performance, risk, and reputation.

However, all stakeholders would have to engage with the single measure, and this is a challenge when there is plurality in how performance might be measured. An innovative approach has been the use of composite indicators, and when indexes are developed, there is then a derived aggregated value (Singh et al. 2009). Chen and Delmas (2011) argue that the multidimensionality of the CSP construct is the primary difficulty in measuring CSP, and this plurality of factors cannot be aggregated easily into a single overarching, or headline, measure. Further a single measure may not allow for sufficient clarity on the distance between those organizations that are promoting social responsibility and those who are not. Aggregated scores lack comparability and interpretability characteristics when scores are aggregated to a single parameter especially when hard and soft criteria are amalgamated into one weighted score (Chen and Delmas 2011). Singh et al. (2009) identify that normalization and weighting of indicators are generally associated with subjective measurements and show a high level of arbitrariness and that with aggregation there need to be scientific rules that guarantee consistency and meaningfulness of composite indexes. This challenge between subjectivity and objectivity is reflected especially in the multidimensionality of composite indexes that aggregate both soft and hard elements (Singh et al. 2009).

CSP also has cultural dimensions such as national culture, geographic region, and level of economic development (Ho and Wang 2012). The research concludes that higher levels of CSP are associated with higher power distance between executives and staff, more collectivist societies, more masculine tendencies, and more uncertainty avoidance. Comparison of the CSR content of annual reports is not enough as a measurement mechanism (Font et al. 2012), and the use of a single aggregate metric has its limitations. The CSP profiling approach can add value by:

1. Stakeholders defining their own standards
2. Profiling driving outcome-based assessment, i.e., the measures developed are hard in terms of being observable and quantifiable
3. Plural perceptions being disaggregated by stakeholder groups in order to give meaning to each group
4. Identifying areas of trade-offs
5. CSP decision-making within the organization giving a clear typology of the CSR activities undertaken (Mahon and Wartick 2011)

CSP profiling must be sector specific and have salience for all stakeholder groups (Mahon and Wartick 2011). In order to demonstrate "value" to a multiple stakeholder audience, an organization should review its CSP, identify potential areas for improvement, and then communicate this back to their shareholders and other stakeholders (Manning et al. 2006). MNCs that ignore CSP and CSR run the risk of increased government and civil scrutiny and potentially intervention (Mahon and Wartick 2011). This is a consideration with regard to socially responsible investing (SRI), i.e., the practice of choosing investment stocks on the basis of environmental and social screening (adapted from Ziegler and Schröder 2010). This would account for around 10% of current asset funds in the United States and Europe (Ziegler and Schröder 2010). As has been previously discussed, indexes such as the DJSI are widely used as a measure for SRI.

The gap between disclosure and performance is considered by Font et al. (2012), especially the difficulty in objectively quantifying CSP either as an aggregate metric or for individual components. Thus greenwashing may actually be a form of trade-off or hypocrisy, i.e., focusing on one CSP parameter that paints the organization in a good light while failing to report CSP in an area where the reverse is the case. This avoiding of public disclosure or "greenhushing" may be as a result of concern over being seen as a target by either individual or groups of stakeholders. Martens (2008) differentiates between greenwashing, greenhushing, and "green telling" the latter focusing on the brand value associated with green credentials. Greenhushing might be a reaction to the concern of a corporation that it might be seen as greenwashing so that it hides its green credentials to avoid confusion. Conversely, greenhushing as a term could be used to describe the practice of only disclosing some of the organizational impacts that influence CSR performance. It could be argued that this social self-disclosure or the use of soft management claims is a strong context in which strategic ambiguity or greenwashing can take place as self-disclosure may not actually relate to real organizational performance. Self-serving bias exists in the corporate narrative used in environmental disclosures not just in their amount or thematic content (Cho et al. 2010 cited by Font et al. 2012).

7.6   Teaching Business Ethics

In this chapter, the vocabulary and context associated with CSR have been discussed in depth.

There have been many examples of transactional forms of CSR, e.g., the use of SRI, development of CSP metrics, and wider CSP profiling. While of value in terms of quantifying both the hard and soft elements of CSR, this approach lacks traction unless there is a transformational element in terms of the MNCs ongoing internal and external narrative and a proactive element of engagement from senior management. Stakeholders will recognize, despite organizational use of strategic ambiguity, the narrative and behavior associated with organizational greenwashing and greenhushing. Mechanisms to determine the behavior of the managers and executives of today have been discussed; however, consideration needs to be given to how business ethics is taught to the managers and executives of tomorrow.

Some faculties would suggest that the dearth of business ethics in teaching programs is due to the lack of faculty expertise or an unwillingness to develop a more integrated approach across the curricula; thus a stand-alone ethics class should be introduced (Floyd et al. 2013). This is however only one point of view. Another point of view would be that ethics needs to be fully embedded in what a student does in terms of their methods of study, their behavior toward others, etc. This questions the notion of whether ethical behavior can actually be taught and, if it is taught, from which moral or ethical standpoint. This should cause faculties to reflect on what business ethics actually is and how they enact their own social responsibility in their interaction with students. For many students the first business they fully interact with is actually the university that they attend.

Students are expected to recognize and distinguish between these approaches and elements in the secondary literature they engage with on CSR and also to develop their own personal analytical skills over time. This includes the ability to develop knowledge and understanding and be able to explain or discuss key terms through extending their intellectual abilities to evaluate the strength of evidence and to provide cohesive scrutiny of detailed concepts and theories and provide a logically constructed, persuasive narrative that is well presented for the reader. To make sense of the body of literature available, it is not enough to simply reproduce knowledge from others on the subject of CSR. Using the learning itself as a lens, students need to comprehend and in some ways "live" the process through which corporations have both expressed their social responsibility and also demonstrated it, or the lack of it, in practice. They need to understand the evolution of the concept of the responsibility of corporations to a range of stakeholders and then where each organization positions themselves in order to deliver value to those stakeholders sometimes with a clear strategy to deliver value in favor of some stakeholders much more than others.

Friedman (1962) stated, "There is one and only one social responsibility of business – to use its resources and engage in activities designed to increase its profits." Friedman therefore proposes that the greatest benefit to the public good is allowing businesses to operate with minimum regulation in their continued pursuit of profit without considering what is deemed morally correct by wider civic society (Manning 2015a). This is one but not a sole viewpoint. Philips et al. (2003) consider that stakeholder theory is a theory of both organizational management and ethics suggesting "all theories of strategic management have some moral content, though it is often implicit." It is the understanding by learners, of the implicit and explicit elements of CSR that is crucial during the learning process as well as their abilities to differentiate between the two. Students have grown up in a society where "distinctions between right and wrong have become blurred and where unethical behaviour is observed and even expected in high-profile leaders" (Kidwell 2001:45). In light of corporate scandals in the United States (USA) there is renewed interest in how business ethics are taught (Sims and Felton 2006). Moreover the authors consider that educational establishments are reexamining the teaching of business ethics, the related content and pedagogy, and their responsibilities to develop graduates who can go on to be managers who "live lives of integrity and ethical accountability" (Sims and Felton 2006:297). They suggest that four questions need to be answered in the development of any course:

1. What are the objectives or targeted learning outcomes of the course?
2. What kind of learning environment should be created?
3. What learning processes need to be employed to achieve goals?
4. What are the roles of the participants in the learning experience?

The subject of business ethics should be an integrated core element of business management study with a dual focus on both analytical frameworks and their applications to business disciplines (Gandz and Hayes 1988). This requires the development of suitable materials and appropriate implementation into the pedagogy. The goal of an "experiential learning" approach is to encourage students to reflect upon the complexities of responsible business education in authentic business contexts (Murphy et al. 2012 citing Herrington and Herrington 2006). They propose a range of pedagogies that promote reflection including internships, practical projects, case studies, group work, and observing and participating in artistic performances or cultural events. The themes outlined in this chapter are key to the student's understanding of wider human interaction with the concept of social responsibility and comprehending to how organizations develop strategy in this area. Sims and Felton's four questions are now considered in turn:

1. *What are the objectives or targeted learning outcomes of the course?*

Ten ethical outcomes have been characterized which should be embedded within teaching programs, although this can be achieved in both an implicit and an explicit approach, within single modules, or across a whole program (Floyd et al. 2013). These ethical outcomes are:

(a) Examining the pressures of the current business environment
(b) Identifying the benefits of virtuous conduct in creating wealth
(c) Identifying the consequences of unethical behavior
(d) Establishing a culture that reinforces personal integrity and honesty
(e) Fostering dialogue about ethics and values
(f) Foundations of decision-making
(g) Motivating others to understand values
(h) Creating better systems that monitor conduct and the consequences of dishonesty
(i) Clarifying rules for academic publications, e.g., the use of plagiarism, etc.
(j) Increasing communication between academics and practitioners

Therefore in order for students to understand what is required of them, learning outcomes need to be clearly defined for individual sessions as well as for course modules and the overall course program.

2. *What kind of learning environment should be created?*

Across a given class, students will have different levels of motivation, different attitudes about teaching and learning, and different responses to specific classroom environments and instructional practices (Felder and Brent 2005). The learning environment must be such that learners feel comfortable in engaging with a subject where there is no unique answer and where all points of view expressed within the learning environment could be equally valid, provided that they are supported by appropriate evidence. The tutor needs to facilitate the exploration of a body of literature, which, as already discussed, often does not have a clear consensus and where plurality occurs at many levels. The literature can be explored in a range of ways through case studies, question and answer sessions, and student seminars where they are asked to consider a range of themes or approaches.

Ultimately, learners need to develop their own opinions, their narrative and argument, and in time their own CSR concepts and frameworks that demonstrate their ability to be original and intellectually creative. It is this ability that MNCs seek in their executives and strategic leaders of tomorrow.

3. *What learning processes need to be employed to achieve goals?*

CSR lends itself to an active learning approach; however some students may prefer a more passive experience, so it is important to include both elements in the teaching of business ethics. The literature and the associated case studies are constantly evolving with examples of CSR strategy, CSP, the use of strategic ambiguity, greenwashing, greenhushing, etc. virtually on a daily basis. This evolving landscape of evidence and literature makes the learning environment of business ethics vibrant and rich, so it is important that students are asked to provide their own contemporary examples to demonstrate their understanding of the theory that is being explored. The use of case studies to prompt intellectual creativity and to encourage students to critically reflect is a key part of the learning process.

4. *What are the roles of the participants in the learning experience?*

The facilitation role is especially important. The lecturer is required to guide learning and to assist the students to benefit from using a range of learning styles. Felder and Brent (2005) distinguish between students having a surface approach and a deep approach to learning.

In a surface approach to learning, students require the academic process to be about knowledge transfer, rote learning, and mechanistic regurgitation in order to get the marks required to pass the module or the course at a specific grade in a purely transactional interaction with the material. This type of learning includes little effort to actually understand or critically reflect upon the material being taught.

Conversely a deep approach to learning means that the students will look at the applicability of the concepts and theories being taught, and this should be promoted by the lecturer when engaging with business ethics literature. The student has a responsibility for actively participating in their own learning experience. This can mean that they need to reflect more deeply on their own abilities and not only their strengths but also the gaps in their knowledge or their intellectual skills. Education is about engaging the whole person in a transformative process, and this requires an individual to actively engage with the learning environment and the wider experience of knowledge creation.

7.7 Conclusions

The use of the term CSR in the business narrative is ubiquitous. There is a complex set of definitions and conceptual arguments that seek to describe what CSR is and does in the business context. Corporate dialogue can be interpreted at the primary level alone or considered for the shadow dialogue that can often sit beneath the surface too. Furthermore there is a plurality of expectations as to what CSR delivers to both individual and, a multiplicity of, organizational stakeholders. CSP and the metrics associated with determining individual organizational behavior are an industry construct designed to quantify an organization's CSR profile and thus to compare different MNCs in terms of mode of operation and investment risk. The challenge is the trade-off between quantifying different hard and soft elements of CSR and CSP. Aggregate scoring systems rely on some basic assumptions and an appropriate logic with which the hard and soft elements of CSR are internally weighted. This transactional approach can be influenced by a number of factors and be of value to some stakeholders more than others.

Thus, there may be stakeholder concerns about an organization focusing on one CSP parameter that paints the organization in a good light while failing to report CSP in an area where the reverse is the case. Quantifying organizational ethical behavior is a dynamic process underpinned by a determination of what society believes it is to be ethical, moral, legal, or indeed legitimate.

The business student needs to both understand the principles of and embrace the subject of CSR and all the associated nuances while being aware that there is rarely only one approach or indeed a sole, correct answer to any question posed in the subject area.

References

Andrews KR (1987) The concept of corporate strategy. Richard D Irwin Inc, New York

Armstrong SJ (1977) Social irresponsibility in management. J Bus Res 5:185–213

ASA (n.d.) ASA ruling on Tesco Stores Ltd. 4th September 2013, Available at: https://www.asa.org.uk/Rulings/Adjudications/2013/9/Tesco-Stores-Ltd/SHP_ADJ_224880.aspx#.VellO7SCbzI. Accessed 01 Sept 2015

Barnard CI (1938) The function of the executive. Harvard University Press, Cambridge MA

Bell S, Morse S (2003) Measuring sustainability: learning by doing. Earthscan Publications Ltd, London

Bourlakis M, Maglaras G, Gallear D, Fotopoulos C (2014) Examining sustainability performance in the supply chain: the case of the Greek dairy sector. Ind Mark Manag 43:56–66

Buchann D, Huczynski A (2004) Organizational behaviour: an introductory text, 5th edn. Pearson Education Limited, Madrid

Buchanan DR (2008) Autonomy, paternalism and justice: ethical priorities in public health. Am J Public Health 98(1):15–21

Carroll AB, Shabana KM (2010) The business case for corporate social responsibility: a review of concepts, research and practice. Int J Manag Rev 12(1):85–105

Celnik D, Gillespie L, Lean MEJ (2012) Time-scarcity, ready-meals, ill-health and the obesity epidemic. Trends Food Sci Technol 27:4–11

Chen CM, Delmas M (2011) Measuring corporate social performance: an efficiency perspective. Prod Oper Manag 20(6):789–804

Cho CH, Roberts RW, Patten DM (2010) The language of U.S. corporate environmental disclosure. Acc Organ Soc 35(4):431–443

Craig RJ, Amernic JH (2004) Enron discourse, the rhetoric of a resilient capitalism. Crit Perspect Account 15(6-7):813–852

Crapanzano V (1990) On dialogue. The interpretation of dialogue. Maranhao, University of Chicago Press, Chicago

Delmas M, Burbano VC (2011) The drivers of greenwashing. Calif Manag Rev 54(1):64–87

Drucker PE (1954) The practice of management. Harper and Row Publishers, New York

Du S, Bhattacharya CB, Sen S (2010) Maximizing business returns to corporate social responsibility: the role of corporate social responsibility communication. Int J Manag Rev 12:8–19

Ducassy I (2013) Does corporate social responsibility pay off in times of crisis? An alternate perspective on the relationship between financial and corporate social performance. Corp Soc Responsib Environ Manag 20:157–167

European Commission (2011) Corporate social responsibility: a new definition a new agenda for action. Available at: http://europa.eu/rapid/press-release_MEMO-11-730_en.htm. Accessed on 5 Sept 2015

Eisenberg EM (1984) Ambiguity as strategy in organizational communication. Commun Monogr 51:227–242

Fairclough I, Fairclough N (2012) Political discourse analysis: a method for advanced students. Routledge, Taylor and Francis Group, London

Fassin Y, Van Rossem A, Buelens M (2011) Small-business owner-managers' perceptions of business ethics and CSR concepts. J Bus Ethics 98(3):425–453

Felder RM, Brent B (2005) Understanding student differences. J Eng Edu 94(1):57–72

Floyd LA, Xu F, Caldwell C (2013) Ethical outcomes and business ethics: towards improving business ethics education. J Bus Ethics 117(4):753–776

Font X, Walmsley A, Cogotti S, McCombes L, Häusler N (2012) Corporate social responsibility: the disclosure–performance gap. Tourism Manag 33:1544–1553

Franks JR (2014) Sustainable intensification: a UK perspective. Food Policy 47:71–80

Freeman RE (1984) Strategic management: a stakeholder approach. Pitman Publishing Inc., Marshfield MA

Friedman M (1962) Capitalism and freedom. University of Chicago Press, Chicago

Gandz J, Hayes N (1988) Teaching business ethics. J Bus Ethics 7(9):657–669

Garriga E, Melé D (2004) Corporate social responsibility theories: mapping the territory. J Bus Ethics 53:51–71

Griffith CJ, Livesey KM, Clayton DA (2010) Food safety culture: the evolution of an emerging risk factor? Br Food J 112(4):426–438

Herrington A, Herrington J (eds) (2006) Authentic learning environments in higher education. Information Science Publishing, London

Ho FN, Wang HMD (2012) A global analysis of corporate social responsibility: the effects of cultural and geographic environments. J Bus Ethics 107:423–433

Horner J (2003) Morality, ethics and law: introductory concepts. Semin Speech Lang 24(4):263–274

Iivari N, Abrahamsson P (2002) The interaction between organizational subcultures and User Centred Design – a case study of an implementation effort, Proc 35th Hawaii Int Conference on Systems Science

Johnson L (2003) After Enron: remembering loyalty discourse in corporate law. Delaware J Corporate Law 28(1):27–73

Johnson G (1988) Rethinking incrementalism. Strateg Manage J 9:75–91

Joyner BE, Payne D (2002) Evolution and implementation: a study of values, business ethics and corporate social responsibility. J Bus Ethics 41:297–311

Khatib TM (1996) Organizational culture, subcultures, and organizational commitment. Retrospective Theses and Dissertations. Paper 11540

Kidwell LA (2001) Student honor codes as a tool for teaching professional ethics. J Bus Ethics 29(1-2):45–49

Leitch S, Davenport S (2007) Strategic ambiguity as a discourse practice: the role of keywords in the discourse on sustainable biotechnology. Discourse Stud 9(1):43–61

Lynam JK, Herdt RW (1989) Sense and sustainability: sustainability as an objective in international agricultural research. Agric Econ 3:381–398

Lyon TP, Maxwell JW (2011) Greenwash: corporate environmental disclosure under threat of audit. J Econ Manag Strateg 20(1):3–41

Manning L (2015a) Determining value in the food supply chain. Br Food J 117(11):2649–2663

Manning L (2015b) Health and well-being vulnerability of the socio-economically disadvantaged: the role of food. In: Rayman-Bacchus L, Walsh P (eds) Corporate Responsibility & Sustainable Development: the nexus of private and public interests. Routledge, London

Manning L, Baines RN, Chadd SA (2006) Ethical modelling of the food supply chain. Br Food J 108(5):358–370

Manning L (2013) Corporate and consumer social responsibility in the food supply chain. Br Food J 115(1):2–29

Martens C (2008) Greenhushing..... Schhhhh. From http://www.kommunikationsforum.dk/artikler/greenhushing-schhhhh. Accessed 16 Dec 2013

Maynard H, Nault J (2005) Big farms, small farms: strategies in sustainable agriculture to fit all sizes. http://www.aic.ca/pdf/AIC_2005_ENG.pdf. Accessed 4 Dec 2014

McGreogor A (2004) Sustainable development and "warm fuzzy feelings": discourse and nature within Australian environmental imaginaries. Geoforum 25(5):593–606

Mahon J, Wartick SL (2012) Corporate social performance profiling: using multiple stakeholder perceptions to assess a corporate reputation. J of Public Affairs 12(1):12–28.

Moroney R, Wundsor C, Aw YT (2011) Evidence of assurance enhancing the quality of voluntary environmental disclosures: an empirical analysis. Account Finance 52(3):903–939

Murphy R, Sharma N, Moon J (2012) Empowering students to engage with responsible business thinking and practice. Bus Prof Ethics J 31(2):313–330

Orlitzky M, Louche C, Gond JP, Chapple W (2015) Unpacking the drivers of corporate social performance: a multilevel, multistakeholder, and multimethod analysis. J Bus Ethics:1–20

Pearce CL, Manz CC, Sims HP Jr (2008) The roles of vertical and shared leadership in the enactment of executive corruption: implications for research and practice. Leadersh Q 19:353–359

Philips R, Freeman RE, Wicks AC (2003) What stakeholder theory is not. Bus Ethics Q 13(4):479–502

Pretty J (1994) Alternative systems of inquiry for a sustainable agriculture. IDS Bull 25(2):37–49

Putrevu S, McGuire J, Siegel DS, Smith DM (2012) Corporate social responsibility, irresponsibility, and corruption: Introduction to the special section. J Bus Res 65:1618–1621

Russell, J, and Greenhalgh, T (2009) Rhetoric, evidence and policymaking: a case study of priority setting in primary care. Final report of a research project funded by the UCL Leverhulme-ESRC programme on Evidence, Inference and Enquiry. Open Learning Unit. Res Dept Primary Care & Population Health. UCL

Scherer AG, Palazzo G (2011) The new political role of business in a globalized world: a review of a new perspective on CSR and its implications for the firm, governance and democracy. J Manag Stud 48(4):899–931

Seedhouse D (1988) Ethics: the heart of healthcare. John Wiley & Sons, Chichester

Selznick P (1957) Leadership in administration. Harper and Row Publishers, New York

Sim AB, Fernando M (2010) Strategic ambiguity and ethical actions. Oxford Business & Economics Conference Oxford, UK: Oxford University p 1–23

Sims RR, Felton EL Jr (2006) Designing and delivering business ethics teaching and learning. J Bus Ethics 63(3):297–312

Simon HA (1945) Administrate behavior. Free Press, New York

Smith SW, Atkin CK, Roznowski J (2006) Are "drink responsibly" alcohol campaigns strategically ambiguous. Health Commun 20(1):1–11

Singh RK, Murty HR, Gupta SK, Dikshit AK (2009) An overview of sustainability assessment methodologies. Ecol Indic 9:189–212

Stigliz JE (2006) Making globalization work. Allen Lane, London

The Grocer (2013) Tesco horsemeat ad misled consumers, ASA rules Available at: http://www.thegrocer.co.uk/channels/supermarkets/tesco/tesco-horsemeat-ad-misled-consumers-asa-rules/349090.article. Accessed 1 Sept 2015

Torugsa N, O'Donohue W, Hecker R (2012) Capabilities, proactive CSR and performance in SMEs: empirical evidence from an Australian manufacturing industry sector. J Bus Ethics 115(2):383–402

Van de Ven A, Schomaker M (2002) The rhetoric of evidence-based medicine. Health Care Manag Rev 27:89–91

Van Marrewijk M (2003) Concepts and definitions of CSR and corporate sustainability: between agency and communion. J Bus Ethics 44:95–105

Van Stekelenburg A, Georgakopoulos G, Sotiropoulou V, Vasileiou KZ, Vlachos I (2015) The relationship between sustainability performance and stock market returns: an empirical analysis of the Dow Jones Sustainability Index, Europe. Int J Econ Finance 7(7):74–87

Walton D (2007) Media argumentation. Cambridge University Press, Cambridge

Wexler MN (2009) Strategic ambiguity in emergent coalitions: the triple bottom line. Corp Commun Int J 14(1):62–77

World Commission on Environment and Development (1987) Our common future. Oxford University Press, Oxford

Wood D (2010) Measuring corporate social performance: a review. Int J Manag Rev 12:50–84

Wood DJ, Jones RE (1995) Stakeholder mismatching: a theoretical problem in empirical research on corporate social performance. Int J Organ Anal 3:229–267

World Business Council for Sustainable Development (WBSCD) (2003) Cross-cutting themes. Available at: http://oldwww.wbcsd.org/DocRoot/7ApjAG0YjGBKx83eok6O/cross-cutting. pdf. Accessed on 10 Sept 15

Ziegler A, Schröder M (2010) What determines the inclusion in a sustainability stock index? A panel data analysis for European firms. Ecol Econ 69:848–856

Chapter 8
Whistleblowing: Food Safety and Fraud

Yasmine Motarjemi

8.1 Introduction

The Council of Europe defines a whistleblower as "any person who reports or discloses information on a threat or harm to the public interest in the context of their work-based relationship, whether public or private" (COE 2014). The term "reports" refers to internal reporting within an organization or enterprise, while the term "discloses" refers to reporting to an outside authority or to the public.

I heard the term "whistleblowing" for the first time sometime in the early 2000s when I was working as the food safety manager in a multinational food company. I remember that at a weekly department meeting, the director of the department shared an article on the subject and asked rhetorically, "I wonder if we are blowing the whistle often enough?" At that time, I did not know that 1 day, I would end up as a whistleblower in the company. It is this professional and personal experience that has prompted me to write this article. However, this is not about my case, but the bigger issue of whistleblowing and what it means for society.

In recent years, with the revelations of Bradley Manning and lately those of Edward Snowden, whistleblowing has become controversial because it is alleged that national security or interests have been compromised. However, the phenomenon is not new, and there have always been individuals who have gone against widely held beliefs to reveal information of critical importance to society. Although in the early days they were not seen as whistleblowers, they were, like Cassandra, not always appreciated or heard. For certain individuals, the term whistleblower has a negative connotation (e.g., a snitch or tattletale); yet, most whistleblowers have

Motarjemi Y. Whistleblowing: Food Safety and Fraud. 2015; Food Safety Magazine, (July 2014):1–12. Reprinted by permission of Food Safety Magazine.

Y. Motarjemi (✉)
Consultant, Rue de la Porcelaine, 10; CH-1260, Nyon, Switzerland
e-mail: yasmine.motarjemi@bluewin.ch

© Springer International Publishing AG 2018   147
R. Costa, P. Pittia (eds.), *Food Ethics Education*, Integrating Food Science and
Engineering Knowledge Into the Food Chain 13,
https://doi.org/10.1007/978-3-319-64738-8_8

high ethical and moral characters, and many have suffered great mental, physical, and economic hardships to render this service to society.

In the area of public health, one notable early whistleblower was Ignaz Semmelweis (1818–1865), a physician working in Vienna. I learned of his story when I started working at the World Health Organization (WHO) as a food safety scientist in the early 1990s. Semmelweis had recognized that the high maternal mortality rate in Viennese hospitals due to puerperal fever was caused by the lack of handwashing by doctors who had previously performed autopsies. He even discovered an effective intervention of washing hands in carbolic acid (phenol). However, his insight was ignored, perhaps because his peers were resistant to change or simply disliked criticism. This story was told to me by the then director of the department of food safety at the WHO, Dr. Fritz Käferstein, who compared it to the situation of food safety that had yet not received the recognition that it has today. Back then, even WHO member states and donor agencies were not very supportive of the nascent food safety program. Infant diarrhea and, generally, diarrheal infections, such as cholera, were attributed to contaminated water, but not food (Motarjemi et al. 2012; Motarjemi et al. 1993). So each time we were confronted with the lack of appreciation for food safety by our fellow public health colleagues, Fritz Käferstein would cite Semmelweis's story.

Although, at that time, we did not see our efforts of alerting and campaigning for food safety as "whistleblowing," in hindsight, we were also on some kind of whistleblowing journey. Despite our continuous attempts to draw attention to the scientific evidence, food safety remained an afterthought at best. Unfortunately, it required a succession of food safety crises (bovine spongiform encephalopathy, dioxins, deadly foodborne disease outbreaks, such as *Escherichia coli* O157 infections) and resulting trade disruptions to bring about a radical change in the public perception and a realization by governments of the importance of food safety to health and to food supply.

8.2 Misperceptions

There are different reasons for whistleblowers to be negatively perceived:

1. Some individuals have obtained their information through illegal means, like a hacker who steals data.
2. The information they reveal may undermine national security or interests.
3. Some whistleblowers are motivated by revenge against an employer or by personal gain.
4. Whistleblowing may be reminiscent of political denunciations and collaboration with repressive states.

There may also be psychological reasons for feeling resentment toward whistleblowers. For instance, everyone probably feels some degree of uneasiness at the thought of being exposed for a transgression of the law or moral values, however minor. Such feelings are possibly a projection of our own inner fears.

Also, some people perceive a whistleblower as someone who disturbs their peace of mind with a truth, that is, a reality that makes them uncomfortable. Colleagues of a whistleblower may be torn between fear of compromising their own situation and feelings of cowardice and guilt for not supporting the whistleblower.

As the French mathematician and philosopher Blaise Pascal (1623–1662) stated, "As men are not able to fight against death, misery, ignorance, they have taken it into their heads, in order to be happy, not to think of them at all."[1] A situation is often represented by the three wise monkeys embodying the principle of "see no evil, hear no evil, speak no evil."

Whatever the reason, the reality is different. In fact, not only do whistleblowers render a great service to society but also they often do it at the price of a huge personal sacrifice. Once, an officer working in a governmental organization who learned about my story as a whistleblower told me, "Lady, you are paying a high price for letting us learn the truth." Therefore, the courage and sacrifice of whistleblowers should be valued and praised rather than denigrated and despised. Most importantly, it is the message rather than the messenger that should be the focus of the employer and, if that is not the case, at least of the responsible regulatory authorities.

Whistleblowing is and should be seen as a civic action. A true whistleblower is motivated by moral purposes and professional integrity, and whistleblowing should not be denigrated because of the ill-perceived actions of a few. Considering today's globalized food supply, illegal behavior, reckless risk taking, or willful negligence can take on huge health and trade dimensions, as experienced with the melamine adulteration of milk powder and the horse meat scandal. Whistleblowing provides an important approach in meeting the daunting challenges of food safety in modern society (Motarjemi and Lelieveld 2014a). Against whistleblowing is perhaps one of the most important lines of defense.

8.3   Regulations

In recognition of the above, many countries are introducing laws and regulations to encourage and protect whistleblowers from unfair treatment by their employers. Some countries, such as the USA, even provide whistleblowers with financial incentives in cases of significant economic fraud. Regrettably, these do not apply to other values of society, for example, health, environment, human, and animal well-being.

The Council of Europe has prepared a recommendation on the subject. With regard to protection of whistleblowers, it stipulates that "whistleblowers should be protected against retaliation of any form, whether directly or indirectly, by their employer and by persons working for or acting on behalf of the employer." Such

---

[1] Translation from the French: "Les hommes n'ayant pu guérir la mort, la misère, l'ignorance, ils se sont avisés—pour se rendre heureux—de n'y point penser."

retaliation might include dismissal, suspension, demotion, loss of promotion opportunities, punitive transfers and reductions in or deductions of wages, harassment, or other punitive or discriminatory treatment (Motarjemi and Lelieveld 2014a).

In the UK, the Public Interest Disclosure Act (1998) protects workers from detrimental treatment or victimization from their employer if, in the public interest, they expose wrongdoing (FSA 1998) (UK Food Standards Agency). In implementing the act, the UK Food Standards Agency has extended the protections to workers in the food industry, whether or not the information is confidential and whether or not the wrongdoing occurs in the UK. Qualifying disclosures include a criminal offense, the breach of a legal obligation, a miscarriage of justice, a danger to the health and safety of any individual, damage to the environment, and deliberate concealment of information related to any of the aforementioned five matters.

In the USA, a series of laws has been enacted to protect employees who blow the whistle on food safety violations. For instance, under the US Food and Drug Administration, the Food Safety Modernization Act of 2011 has provisions against retaliation toward whistleblowers by food businesses (FDA 2011; OSHA 2014).

France has also developed a number of regulations to protect whistleblowers in relation to corruption, as well as for public health and safety. Among these are Articles 2013–316 of the Code of Labour (2013) relating to the independence in scientific expertise in public health and environment, as well as Article L 4133–1 for protection of whistleblowers in businesses (Légifrance 2016). In 2016, with the adoption of « La Loi Sapin II ", France strengthened its legislation for the protection of whistleblowers.

Since 2008, Switzerland is also in the process of regulating whistleblowing. However, its draft law met major opposition in 2015, as it aimed to protect the reputation of businesses rather than protecting whistleblowers (Motarjemi, Y. Glick, A. (2014 and 2015); Motarjemi (2015).

There are also a number of other countries (e.g., Luxembourg, Slovenia, and Hungary) that have legislation for protection of whistleblowers, but some are more limited in scope to anticorruption, or they do not have in place an infrastructure to handle such complaints.

8.4 The Impasse of Whistleblowers

In some countries, regulations require that employees should report their observations first internally to their own management and, in case there is no follow-up or satisfactory response, to report their concerns to regulatory authorities. However, a common problem for whistleblowers is that employers often ignore the reports and do not follow up the issue. Instead, they subject the whistleblower to retaliatory measures, such as psychological harassment, transfer, or dismissal. At times, even regulatory authorities fail to investigate. This was my personal experience.

Another difficulty is that the whistleblower may be obligated to report to the very person(s) responsible for the failure. Such a situation inevitably leads to retaliatory measures to silence the whistleblower; this is particularly a problem if a senior manager is involved. Also, under present workplace conditions, a whistleblower typically has to assess the importance of a wrongdoing alone without any outside support. This also means that the whistleblower assumes the consequences of reporting the events. Where colleagues also are aware of the situation yet remain silent, the whistleblower may be too intimidated to report, out of fear that he/she may be misjudging the risk or the importance of the wrongdoing, or there may be another hidden or misunderstood explanation for his/her observations (Dehn and Callan 2004).

To encourage employees to come forward with their observations, laws for protection of whistleblowers should consider the risks and consequences for employees and include effective sanctions against employers who retaliate. Also, governments should provide legal assistance for the employees to help them take their case to the courts of justice, where necessary. Furthermore, people who suspect a wrongdoing but cannot provide direct evidence for their concern, or whose information cannot be validated, should not be penalized in any way for raising the issue, particularly if this takes place in the workplace.

8.5 Application to Food Safety and Risk Management

Since ancient times, food fraud (sometimes referred to as economically motivated adulteration) has been a concern. Although motivated by financial gain, this sometimes impacts the safety of products. Recent examples of adulteration are chili with the carcinogen Sudan Red, sunflower oil with mineral oil, and milk with melamine (Motarjemi 2014a). Importantly, with the increased international trade in food and the globalization of the food market, these events have taken a much broader dimension; when they occur, the consequences can be far-reaching and devastating. For instance, in 2008 melamine was used to mask the adulteration of milk in China, and infant formula made from the contaminated milk resulted in kidney damage in over 300,000 infants, with 54,000 of these infants hospitalized and 13 died. In 2012, Jiang Weisuo, the man who first alerted authorities to what would become the melamine-tainted milk scandal, was murdered in the city of Xi'an (Reason 1995).

Detecting unpredictable fraudulent practices is almost impossible through conventional approaches, such as product testing; this further highlights the importance of whistleblowing. However, there are other reasons that underscore the need for whistleblowing. One is the corruption of the systems meant to ensure the safety of food products. For example, extra bonuses or promotions are given in exchange for silence and not reporting food safety problems to management. Another reason has to do with structural deficiencies, for instance, when auditors (internal or external) are in the position of a conflict of interest and subsequently downplay deficiencies or turn a blind eye to gaps or weaknesses of a system they are meant to review. Scientific biases and conflicts of interest are also concerns with experts involved with the risk assessment of biological and chemical hazards in food or technologies used to produce foods.

Some structural and organizational deficiencies may be difficult to characterize as a public health threat and henceforth to denounce, the reason being that their consequences for the safety of products may not be immediate, but rather more long-term in nature, and the prospect of an adverse event happening may not be definite. Examples of such situations are appointments made on the basis of nepotism rather than professional skill or experience, staff working under unrealistic time frames or under duress, neglect in training personnel for their job and/or insufficiently supervising them, downplaying deficiencies, carelessness or inconsistency in communication, or in general having unresponsive or slow management systems. Such deficiencies are referred to as "latent failures" (Reason 1995; Gerald Moy 2014; Motarjemi 2014b).

A company's culture based on fear, and which discourages reporting and/or fails to follow up internal reports, constitutes perhaps one of the worst kinds of latent failures. It deprives an organization of opportunities to anticipate adverse events and to take early actions to nip the risks of accidents in the bud. Such situations have been the root cause of many serious accidents in the food industry and others, for example, Snow Brand, Toyota, and British Petroleum (Motarjemi 2014b; Reason 1997). In Switzerland, Nestlé openly acknowledged this management problem. In its book, *Transformational Challenge: Nestlé 1995–2005* (Pfiffner and Renk 2007), the following quote appears: "The unwillingness to report negative events fully and swiftly up the chain of command may be a vestige of the past culture at Nestlé, a culture in which admitting mistakes was not exactly good for your career, and in which internal criticism was "not the done thing."

The culture of learning from mistakes is not yet as widespread as it is in the aviation industry, where even the smallest incident is analyzed and evaluated to prevent repetitions.

To increase profits and create value for shareholders, some companies may cut back on expenditure and investments in food safety, as the added value of such investments is not always visible to consumers and does not constitute a selling point. Such decisions lead to increased risk of organizational failures. A case in point is the policy of a well-known food company to link the bonuses of its managers to a lack of incidents and product recalls, thereby discouraging its managers from reporting incidents or recalling contaminated products. Financial crises may of course exacerbate the situation.

8.6 Whistleblowing: The Backbone of Risk Management

With the extensive industrialization and commercialization of the food supply, the resources of government authorities will never be sufficient to control the safety of the many food operations and products on the market. Also, end-product testing of products, as a sole measure, can in no way be an effective approach for ensuring food safety for detecting and preventing unknown substances that malevolent people may add to products. Therefore, the trust that we can have in food safety depends very much on the following:

- Competence and ethics of professionals working in the food industry
- Liberty and authority given to the staff to report deficiencies or unethical practices internally, or to authorities, without being subject to retaliation and punitive measures
- Commitment by management to address and follow up on reported food safety issues, including structural problems
- Vigilance of food safety authorities in following up and investigating the root cause of deficiencies and incidents up to the highest level of company management

8.7 The Way Forward

The above demonstrates the importance of considering the human factor in food safety and risk management. Although a great proportion of employees are reliable and deserve trust and respect, this cannot be generalized. The scale that the horse meat fraud took before it was actually detected illustrates the point. Therefore, a national system of food safety management needs to be supported by regulations that achieve the following:

- Make the senior directors of a company, such as the CEO, directly accountable for investigating internal reports and taking appropriate measures.
- Severely sanction managers of companies who try to block or do not follow up on internal reports and/or take retaliatory measures (psychological harassment) against those who report failures, deficiencies, or malpractices.
- Protect whistleblowers from civil and legal suits for disclosing public interest information.

Additionally, there is a need to provide advice for those whistleblowers who are unsure whether or how to raise a public interest concern. Those who are subjected to retaliatory measures would also need legal assistance and other types of support.

Where a case relates to issues of international interest, the whistleblower should be enabled to take his/her case to international judicial authorities directly without having to go through a national system. A whistleblower who is a victim of retaliatory measures will rarely have the means (time, energy, funds) to go stepwise through the extensive procedures of a national system, particularly if the national judicial system is slow and impeded by powerful multinational businesses/employers with almost infinite resources and power to influence the national system.

Also, as experienced in the case of Edward Snowden, a national legal system, which is itself under scrutiny as a result of a disclosure, is unlikely to fully operate in an unbiased manner, as most governments naturally give priority to their own national interests. Therefore, in such cases, which are likely to increase in light of the increasingly globalized systems of trade, finances, and communications, the fundamental question of conflict of interest will have to be taken into account. The smaller the country, the more vulnerable it will be to the influence and the power of multinational companies.

8.8 Conclusions

In conclusion, we need to move from merely authorizing whistleblowing to facilitating it so that employees not only dare to come forward with their information, but consider it their moral obligation to do so. In companies where psychological harassment and a culture of fear are exercised, and in countries where there is no protection for whistleblowers, there is little incentive for potential whistleblowers to disclose their concerns, either internally or externally.

Psychological harassment and other retaliatory measures are barriers to whistleblowing. When exercised on an employee, they will have a chilling effect on anyone else who might become aware of a wrongdoing or of a serious food safety issue, and the company will miss the opportunity to control operational risks or improve its system before a serious incident occurs. Failure to remedy this situation comes at the cost of undermining public health, the environment, human rights, and social welfare. It will also foster ideal conditions for corruption to thrive. In addition to endangering public health, the cost to the food industry is also significant as the loss of consumer confidence in the food supply will have a detrimental effect on the food industry as a whole. Unless serious efforts are made to address the problems of communication and accountability with respect to food safety and other such fundamental public interest issues, the health, social, and economic crises that have been observed in the past will continue to occur with all too frequent regularity.

As concluding remarks, I would like to add that throughout my professional career, I have contributed to various scientific and technical aspects of food safety and its management at the international level. I have produced numerous publications and recently two major reference works (Motarjemi and Lelieveld, 2014b; Motarjemi et al. 2014). Yet today, I consider that my biggest contribution to food safety has been my actions as a whistleblower and reporting my concerns regarding the management of food safety, both internally in the company for which I worked and publicly.

When I started my work in WHO as a scientist, one of the key points that I learned was the importance of the human factor in food safety management. At that time, my focus was on consumers and consumer practice. However, through my experience in industry, I realized the crucial and pivotal role of employees, from the CEO down to the worker on the line.

I learned that too often company policies are merely statements of good intentions without always a serious plan for implementing them. I learned that the management may even violate its own policies, a behavior which sets a very negative model for the entire company and fosters a culture of complacency. It gives the message that integrity does not matter and puts in motion opportunities for future failures. I learned that in spite of written policies, in some companies or organizations, whistleblowing is still unwelcome, particularly when the interests of the management itself are engaged. As a consequence, critical information pertinent to health and safety may not be revealed. Large food businesses are typically run by businessmen who have a secondary interest in consumer health and nutrition, and professionals trained in food safety are not always those who win the day in key decisions.

Based on my personal experience, those with humanity and concern for their colleagues or fellow citizens are ejected from the system or, at best, remain at the bottom of the pyramid of hierarchy. Those who are the most callous and lacking compassion are moved upward in the chain of command. However, the hardest lesson was to realize that those who should be the guardian of public health and who should verify the information put forward by whistleblowers, that is, the regulatory authorities, turn a blind eye and ignore the concerns of the whistleblower. Even worse, in some countries, they enact legislation to oblige employees to be silent, which to a person with moral values is most painful and inhumane; moreover, a whistleblower is at risk of becoming an outlaw.

Predictably, the media is more interested to report on wrongdoers, such as Bernard Madoff[2], rather than a whistleblower who sacrifices his/her personal interests, livelihood, and even his/her life for the well-being of society, as if violence, greed, and malfeasance were more gripping than honesty and integrity.

Another disappointing experience has been the apathy of civil societies and their lack of support for whistleblowers. This vacuum of counterforce in the society leaves the well-being of people at the mercy of unscrupulous individuals. It is a lesson that societies have long known but, for some reason, keep forgetting. To wit, "The price of apathy towards public affairs is to be ruled by evil men" (Plato).

Acknowledgments I am grateful to all the colleagues who have reviewed this article and provided valuable comments. Specific thanks to Gerald Moy (retired WHO food safety scientist) and Anna Myers (coordinator of Whistleblowing International Network) for their extensive input in the preparation of this article.

References

COE (2014) Protecting whistleblowers. Council of Europe/European Committee on Legal Cooperation. Retrieved 6 May 2016, from http://www.coe.int/en/web/cdcj/activities/protecting-whistleblowers

Dehn G, Callan R (2004) ODAC. www.u4.no/recommended-reading/whistleblowing-the-state-of-the-art-the-role-of-the-individual-organisations-the-state-the-media-the-law-and-civil-society/

FDA (2011) FDA Food Safety Modernization Act, Section 402. Retrieved 6 May 2016, from https://www.whistleblowers.gov/acts/fda_402.html

FSA (1998) Whistleblowing. Food Standards Agency. Retrieved 6 May 2016, from https://www.food.gov.uk/enforcement/the-national-food-crime-unit/foodfraud/whistleblowing

Légifrance (2016) Légifrance, le service public de l'accès au droit. Retrieved 6 May 2016, from https://www.legifrance.gouv.fr/

---

[2] Bernard Lawrence Madoff is an American convicted of fraud and a former stockbroker, investment adviser, and financier. He is the former nonexecutive chairman of the NASDAQ stock market and the admitted operator of a Ponzi scheme that is considered to be the largest financial fraud in the US history.

Motarjemi Y (2014a) Crisis management. In: Motarjemi Y, Lelieveld H (eds) Food safety management: a practical guide for the food industry. Academic Press, Waltham, MA

Motarjemi Y (2014b) Human factors in food safety management. In: Motarjemi Y, Lelieveld H (eds)
Food safety management: a practical guide for the food industry. Academic Press, Waltham, MA

Motarjemi Y, Lelieveld H (2014a) Fundamentals in management of food safety in the industrial
setting: challenges and outlook of the 21st century. In: Motarjemi Y, Lelieveld H (eds) Food
safety management: a practical guide for the food industry. Academic Press, Waltham, MA

Motarjemi Y, Lelieveld H (eds) (2014b) Food safety management: a practical guide for the food
industry. Academic Press, Waltham, MA

Motarjemi Y, Käferstein F, Moy G, Quevedo F (1993) Contaminated weaning food: a major risk
factor for diarrhea and associated malnutrition. Bull World Health Organ 71(1):79–92

Motarjemi Y, Steffen R, Binder JH (2012) Preventive strategy against infectious diarrhea—a holistic approach. Gastroenterol 143(3):516–519

Motarjemi Y, Moy G, Todd E (eds) (2014) Encyclopedia of food safety. Academic Press, Waltham,
MA

Motarjemi, Y. (2015) The real problem with whistleblowing. LeNews. 19 February 2015 http://
lenews.ch/2015/02/19/the-real-problem-with-whistleblowing/

Motarjemi, Y. and Glick,A. (2014) Switzerland to silence whistleblowers. LeNews. 9 October
2014 http://lenews.ch/2014/10/09/switzerland-to-silence-whistle-blowers/

Motarjemi, Y. Glick, A. (2015) Switzerland could now lead the world on whistleblowing rules.
LeNews 5 May 2015 http://lenews.ch/2015/05/15/switzerland-could-now-lead-the-world-on-whis-
tleblowing-rules/

OSHA (2014) Procedures for handling retaliation complaints Under Section 402 of the FDA
Food Safety Modernization Act – 79. Occupational Safety and Health Administration,
pp. 8619–8632. Retrieved 6 May 2016, from. https://www.osha.gov/pls/oshaweb/owadisp.
show_document?p_table=FEDERAL_REGISTER&p_id=24284

Pfiffner A, Renk H-J (2007) Resource nestle annual report. Retrieved 6 May 2016, from https://
pt.scribd.com/doc/7125649/Resource-Nestle-Annual-Report

Reason J (1995) Understanding adverse events: human factors. Qual Health Care 4(2):80–89.
Retrieved from http://www.ncbi.nlm.nih.gov/pmc/articles/PMC1055294/

Reason JT (1997) Managing the risks of organizational accidents. Ashgate UK, Aldershot

Chapter 9
Communicating Food Safety: Ethical Issues in Risk Communication

Merve Yavuz-Duzgun, Umit Altuntas, Mine Gultekin-Ozguven, and Beraat Ozcelik

9.1 Introduction

All of our lives are influenced by food, and this makes food safety a sensitive issue that gains the interest of all community (Anderson 2000). The concept of food safety has developed by different ways until today.

Early laws related to food safety were based on prohibition while in recent times prevention was introduced. Today, it is well known that risk analysis is an essential tool to produce safe foods and protect public health.

A risk analysis that underlies the EU food safety policy has three parts which are risk assessment, risk management, and risk communication. The Food and Agriculture Organization (FAO) and World Health Organization (WHO) defined these components in three joint expert consultations (Jensen and Sandøe 2002) as follows:

- *Risk assessment* is described as "the scientific evaluation of known or potential adverse health effects resulting from human exposure to foodborne hazards."
- *Risk management* means "the process of weighing policy alternatives to accept, minimize or reduce assessed risks and to select and implement options."
- *Risk communication* is "an interactive process of exchange of information and opinion on risk among risk assessors, risk managers, and other interested parties" (WHO 1995). In other words, risk communication is defined as "the flow of information and risk evaluations back and forth between academic experts, regulatory practitioners, interest groups, and the general public" (Lofstedt 2006). Risk communication differs from risk assessment with regard to regulations.

M. Yavuz-Duzgun • U. Altuntas • M. Gultekin-Ozguven • B. Ozcelik (✉)
Istanbul Technical University, Department of Food Engineering,
Ayazaga Campus Maslak, Istanbul, Turkey
e-mail: ozcelik@itu.edu.tr

© Springer International Publishing AG 2018
R. Costa, P. Pittia (eds.), *Food Ethics Education*, Integrating Food Science and
Engineering Knowledge Into the Food Chain 13,
https://doi.org/10.1007/978-3-319-64738-8_9

Risk assessment is the evaluation of the adverse health effects of food-borne hazards, and the risk assessors have the objective of reducing the risk of these hazards, whereas the regulations concerning risk communication have its own liberality. The control system should protect the consumers and decrease the risk of the food-borne hazards, while consumers have their own right to choose what is beneficial on the market (Jensen and Sandøe 2002).

The four stages of any risk analysis are listed below (Renn 2005):

(a) The severity of the risk which is related to scientific risk assessment.
(b) Acceptability of the assessed risk which means whether it is tolerable or not. This is related to risk evaluation.
(c) Searching the possibilities which could reduce or remove the risk which is related to risk management.
(d) The way of achievement of the transparency, understanding, and agreement to the planned risk management which is related to risk communication.

Consumers are increasingly demanding food quality and safety and want to have more information about their foods such as the origin, ingredients, the environmental, ethical, and technological conditions of food, and all information related to food safety (Verbeke 2005). In the industrialized world, many factors have affected the food safety perception, and this effectively shaped the relationship between the consumers and the food industry. In addition, the intense interest of mass media on food quality and safety caused consumers to alter their risk perception and food suppliers to change their behaviours. Risk perception is an important issue to which should be paid attention by the food communicators. It was also indicated that an effective investigation should be conducted about the risk perceptions in public considering the social and cultural effects (Knox 2000). A reform in risk perception requires an effective risk communication system. Food-specific issues such as improvement of the risk communication program associated with microbial risks and the analysis of the relation between food safety and nutrition are aspects of the risk communication discipline. Communication strategies are related to food hazards with high impact on the public, such as pesticide contamination and novel foods (Lofstedt 2006). The attention paid by risk communicators plays the main role in the success of introducing new technologies, such as food irradiation or genetically modified (GM) foods (Knox 2000).

Ethics has significant contributions to risk evaluation and risk management. It promotes a more open and democratic participation of consumers in the food market by respecting their choices and decisions (Sperling 2010). The principles used in the process of decision-making on significance or magnitude of a risk are crucial. Some ethical questions arise at this point: Who decides a risk is significant? Is the decision-making process based on the scientific principles, or are the values of the society taken into consideration? What is the amount of information to be released? When is the information released? Who decides which part of information will be disseminated and to whom will the information be given? Food culture, public institutions, public movements, policies, and the law have an impact on the responses to these questions. In addition, the convenience of using persuasion and how the public danger relates to private interests are also important issues Lundgren and McMakin (2013) separated all this issues about risk communication into three areas, namely, social ethics, organizational ethics, and personal ethics.

In this chapter, different aspects of risk communication will be firstly explained, and a brief information of the three areas of ethics in risk communication will be given. Secondly, the two tools of risk communication, which are media and labeling, will be explained with regard to the relationship between the ethics and the risk communication.

9.2 Ethics in Risk Communication

9.2.1 Concepts of Risk Communication

Different concepts have been developed from the studies on food risk communication which are summarized below (Lofstedt 2006):

(a) Communication of uncertainty and the role of transparency: Consumers want to be informed about food risk. According to this concept, people are ready to accept uncertainty related to the scientific process of risk management, and all information about uncertainty should be shared with the public. Transparency of risk analysis process requires public involvement. Public opinion about the risk communication has evolved over the years, and today public wants to be considered in risk analysis process. Creighton (2005) explained this alteration by the changing feelings of society in different eras. In the 1950s, the public could accept a decision after being informed about it. However, in the 1960s and 1970s, public demanded to hear about an issue before the decision was made on it. Since 1980's society has desired to be involved in decision- making process which requires a consensus before the decision is applied. Involvement of public in the process of risk analysis provides an increase of organization's credibility and broader information in understanding the risk. However, it can cause loss of organization's control and requires more time.

 The involvement of public into the different stages of risk analysis (risk assessment, risk evaluation, risk management, and risk communication) has both advantages and drawbacks. Consideration of the public concerns at the stage of risk assessment has the benefit of preventing both delayed schedule and increased budget. Public involvement in scenario development where the risk is calculated based on the lifestyle and some other factors may cause a wider number of scenarios with an increased time and costs to complete it (Lundgren and McMakin 2013).

(b) Social amplification: The integration of different models of risk perception and risk communication is considered in the theory of social amplification.

"The social amplification of risk is based on the thesis that events pertaining to hazards interact with psychological, social, institutional, and cultural processes in ways that can heighten or attenuate individual and social perceptions of risk and shape risk behaviour." (Kasperson, 1992)

The use of risk idiom by society and experts constitutes one of the subjects of ethics. It is not possible to give the consumer the message of acceptation to eat a food involving a risk factor. Here, the risk refers to an illness that can also be fatal. Consumers are not interested in how much safe the food is; they want to be informed about whether the food is safe or not. The food industry, national governments, and the media have partial responsibility for the failure of misunderstanding of the concept of risk by the public (Anderson 2000). To prevent this misunderstanding, Anderson (2000) suggested that all parties should enter into dialogue and participate in the risk evaluation process. In addition, Renn (2005) stated that risk perception has a significant place in food safety. It requires its actuality whose rules must be observed if risks are to be sufficiently strived and in a consumer-oriented way. Thus, risk management needs new assessment criteria exceeding the ordinary factors of probability and degree of negative impact (Renn 2005).

(c) Stigma: Stigma has come to light after the study of Goffman (1963) and begun to be researched by social scientists. It is a negative feature that dominates a certain interpersonal or food type issue. For example, Californian strawberries were wrongly thought to be contaminated with *Cyclospora cayetanensis*, though the real guilty was Guatemalan raspberries. Nevertheless, the correction was made; most of the people did not hear the truth and stopped consuming Californian strawberries leading to loss of 20–40 million dollars in California strawberry sales (Goffman 1963).

(d) Trust: Public trust in authorities responsible for providing information has a high impact on public reactions. Some incidents happened in the past which rose from public distrust to some individuals and caused policy makers to fail during the implementation of risk communication programs. As a consequence of these, the area of trust has entered to ethics in risk communication as a new dimension.

Public trust can be provided by presenting simple facts in reply to consumer's questions. These questions are important because consumers are part of the risk communication. Renn et al. (2005) analyzed in detail the telephone calls made to authorities or institutes by consumers who ask for assurance on various food "scandals." Most of the consumers did not ask: "What is the level of the risk?", "What type of adverse health effects can be expected?", or "Am I acutely in danger or trouble?". On the contrary, consumers who are apprehensive ask these questions: "What can/should I do?", "How can I protect myself from the hazard?", and "What can I eat without being in danger?". Moreover, people willing to learn for proof of faith asked: "Who tells the truth?", "Whom should I trust to?", and "Can I trust my food retailer?". While consumers expect competent answers, they frequently obtained merely empty statements or popular science versions of the risk assessments.

The concept of public trust is important to gain acceptance of complex food technologies. Rodriguez (2007) stated that it is not enough to inform the public about the scientific facts of new technology. Lay terminology should be used for an effective risk communication. In addition, technical risk estimates are not adequate to develop a risk management policy to be accepted by consumers.

Risk communicators should give a high attention on consumer issues, especially those related to health and safety as well as new food technologies. Therefore, they need to collaborate with health-related organizations to help convey complex knowledge and information in a manner that accepts community trust.

9.2.2   Ethics Dimensions in Risk Communication

There are three areas of ethics in risk communication as follows: social, organizational, and personal ethics.

Social Ethics Social ethics means how the society judges the individual's behaviour. While the social ethics evolves in time, risk communication has also been changing based on evolving society demands. Lundgren and McMakin (2013) explained how the risk communication relates to social ethics by following viewpoints:

- How the society affects the risk communication
- How the risk idiom is used by society
- By whom and when the risk idiom is used
- Whether the all ethnic and social groups are affected by the risk equally
- In the case of misunderstanding of messages in the risk communication, who endures the results?

Organizational Ethics Organizational ethics encompasses how the code of ethics of organizations handles legitimacy of representation, the designation of the first audience, the release of information, and the attitude toward compliance with regulations.

The legitimacy of representation is an issue both for an organization and the audience. Ethical issues related to the legitimacy of representation include who will be allowed to represent and whether the presented information reflects the truth about the risk. Designation of primary audience is another main organizational ethical issue, and this requires to consider the following: Which parts of the society are at the most risk? Who are the least informed about the choices of the managing the risk or which parts of the society? Who are most involved in making choices about managing the risk?

The another ethical question is when and how much information will be released. The public wants as much information as possible at earlier processes of the risk analysis. On the other hand, organizations show a tendency to give little information and, likely, too late. This can be justified to ensure the validity of scientific observations before sharing with the public. In addition, experience with popular cases such as irradiated foods, genetically modified foods, and functional foods shows that perceived safety of consumers can decrease strikingly when new information is provided even without scientific evidence (Verbeke 2005).

Personal Ethics Personal ethics is the last ethical issue in risk communication. Risk communicators can use persuasion to force an opinion on the public. The duty of risk communicators is to give the risk decision or to help the decision-makers to take their best decision. Sometimes a dilemma can occur between the personal and organizational ethics. If a risk communicator faces a dilemma with his/her organization about the quantity of disseminated information, or something else, she/he has three choices: applies the word of the organization, resigns from the organization, or finds someone else who can solve the problem (Lundgren and McMakin 2013).

Risk communication aims to inform the consumers and reduce the risk by changing their behaviour upon their decision-making. For any food risk situation which is not severe enough and does not require intervention by the government, risk communication processes are easily handled (French et al. 2005).

9.2.3   Labeling

Food labeling is a crucial tool to decrease risks through better communication and allow consumer autonomy. Labeling ethics encompasses all the aspects associated to the definition of the foods to be labeled, the label format, size, information, and impact, whether it should include ingredients, and also whether it is sufficient to present them as a list or by their weight (Sperling 2010). Some other issues can also be considered in view of the consumer autonomy including the prevention of stigmatization of food technology, the protection of the consumer from being manipulated, and the reduction of adverse effects arising from confusion and misinterpretation of information (Thompson 1997).

Scholderer and Frewer (2003) conducted an experiment to determine attitude change of consumers for GM foods by applying three different information strategies (balanced general information, product-specific information, and conventional product advertising) against a control where no information was provided. The balanced general information included some important facts about food biotechnology with the main proponent and opponent views. The product-specific information brochure was shorter and comprised physical attributes of the product and the relationship between these attributes and the application of gene technology. The conventional product advertising promoted the benefits of the product. The surprising result of the study was that the three information strategies caused a negative effect on consumer preferences on GM foods. They all activated the preexisting attitudes of the participants regardless of the type of the information strategy. On the contrary, in the control group, there was a "genetically modified" disclosure, and it was observed that the label information was less likely to activate preexisting negative attitudes against GM foods (Scholderer and Frewer 2003).

Label information on shelf life, nutritional value, and food additives is required to inform the consumers who have the right to know what is in their food. The United Nations Food and Agriculture Organization (FAO) published a paper which presents interviews during June to August 1999 aiming a description of the views of FAO staff on the ethical issues faced to develop the ethics in FAO. In this paper, some of the staff mentioned that the principal duty is not only the giving of correct information to consumers to make their choice, but also adjusting the quantity of the information has significance. A hindrance of choice is a subject in case of too much information. However, the labeling relating to consumer choice was not a major issue raised in the interviews (Bhardwaj et al. 2003). It was also reported that more information causes more concern to the public, so it is a critical question what the right balance of information is.

Another problem is to consider the same food which is safe and unsafe at different times. Graham (2002) exemplified this problem by saturated fats and trans fats. Only a few years ago, all fats were seen to be bad. Later on, the scientists conducted more studies on the saturated fats, and then the saturated fats were seen to be bad. However, today trans fats are presented as the ones that public should be worried about. As a result, an unclear situation arises which can cause a decrease of public trust on regulators. Accordingly, giving a significant amount of information in the first place to the public may not be wise (Lofstedt 2006).

Consumers express their ethical concerns on products by purchasing foods for their ethical qualities in the areas of environmental, biological, social, or fair trade. Additionally, boycotting the products which possess some unethical properties is another way of expression. Pelsmacker et al. (2005) conducted a study in Belgium to assess the significance of different properties and marketing practices of coffees which are ethically labeled according to the consumer attitudes. The used attributes were the type of ethical issue, the amount of information, the label issuer, the distribution and the promotion of strategy, and branding. Distribution strategy was the most important attribute according to the consumers which mean that ethical products should be found in ordinary supermarkets rather than being available only in specialty stores and put up for sale along with nonethical ones. Ethical labeling comes after the distribution strategy in importance ranking. Fair trade-labeled coffee was the most preferred followed by socio-labeled products. Consumers attached less importance to eco- and bio-labels than the fair trade and socio-labels. They care about the issuer of the label in the third place. European government and nongovernmental organizations were preferred more than the national (Belgian) government as label issuers. The study revealed as another important point that the different sociodemographic groups gave the similar scores for the importance of various attributes. It was also indicated in this study that consumers often express a willingness to buy ethical products while these ethically labeled products have low market shares. The reason behind this is probably that most consumers worried more about several product attributes jointly such as price, quality, availability, brand knowledge, attitude, and possibly ethical quality attributes (De Pelsmacker et al. 2005).

Verbeke (2005) studied the reaction of consumers to beef label traceability indications, and they determined that sociodemographic situations make a difference in reactions of consumers. However, in general, the attention to information about traceability and product identification was lower than to the indications on quality. They concluded that the best choice for labels having traceability indications is the inclusion of a quality label accompanied by a single cue referring to traceability (rather than lots of cues and codes such as quality marks and an indication of origin and traceability codes).

9.2.4   Media

Marketing and advertising are useful areas to research consumer behavioral perspective and ensure a development of the food risk communication.

As regards advertisement, 11 billion dollars are being spent by the US industry every year on advertising foods like light beer, sugar drinks, or potato chips. Advertisement budget is supported by heavily funded market research, based on focus groups, interviews, and surveys (Lofstedt 2006).

Media should adopt a responsible approach for a transparent risk analysis during reporting. Any food safety risk should be communicated and not be lost in the column inches and broadcast time, according to Anderson (2000).

It is important that food scientists collaborate with journalists and that their messages should be given in nonscientific language. Moreover, a better relationship between the media and the food industry promotes the confidence of consumers. Risk communicators who can be technical experts, government spokespersons, and "the men in white coats" should engage in an effective dialogue with the media and the public to inform them about the importance of their food handling practices (Griffith et al. 1998).

Media, while giving the food risk message, uses three components: the product, the kind of harm, and the substance, the process, or activity. There is always a scientific uncertainty in the information received by the media. However, the media evaluates this information and redefines it, turning uncertainty in the links between product and compound or substance (e.g., crisps and acrylamide) and harm (e.g., acrylamide and cancer) into certainties. As an example, the issue related to acrylamide in foods in Sweden led a local newspaper to publish the following headline "Acrylamide in crisps causes cancer." Such a question can arise about the goal of the risk message: is the purpose of the message making consumers informed about the risk or is the aim of this changing consumers' attitude leading to not to purchase the food? (Ferreira 2006).

9.3 Conclusions

Food safety risk communication is imperative due to the uncertainty of consumers about food quality and safety.

Media, food scientists, and food industry should collaborate to clarify uncertainty about these issues. The message should be given to consumers in a clear, nonscientific language. As regards the risk communication provided by labeling and media, it is important to be aware of the ethical issues related to these two communication tools.

References

Anderson WA (2000) The future relationship between the media, the food industry and the consumer. Br Med Bull 56(1):254–268

Bhardwaj M, Maekawa F, Niimura Y, Macer DR (2003) Ethics in food and agriculture: views from FAO. Int J Food Sci Technol 38(5):565–577

Creighton JL (2005) An overwiew of public participation. In: Creighton JL (ed) The public participation handbook: making better decisions through citizen involvement, 1st edn. Wiley & Sons Inc., San Francisco, pp 5–27

De Pelsmacker P, Janssens W, Sterckx E, Mielants C (2005) Consumer preferences for the marketing of ethically labelled coffee. Int Mark Rev 22:512–530

Ferreira C (2006) Food information environments: risk communication and advertising imagery. J Risk Res 9(8):851–868

French S, Maule AJ, Mythen G (2005) Soft modelling in risk communication and management: examples in handling food risk. J Oper Res Soc 56(8):879–888

Goffman E (1963) Stigma and social identity. In: Goffman E (ed) Stigma: notes on the management of spoiled identity, 1st edn. Simon Schuster Inc., New York, pp 1–41

Graham M (2002) Democracy by disclosure: the rise of technopopulism. Brookings Institution Press, Washington, D.C

Griffith C, Worsfold D, Mitchell R (1998) Food preparation, risk communication and the consumer. Food Control 9(4):225–232

Jensen KK, Sandøe P (2002) Food safety and ethics: the interplay between science and values. J Agric Environ Ethics 15(3):245–253

Kasperson RE (1992) The social amplification of risk: progress and developing an integrative framework. In: Krimsky S, Golding D (eds) Social theories of risk. Praeger, Westport, pp 153–178

Knox B (2000) Consumer perception and understanding of risk from food. Br Med Bull 56(1):97–109

Lofstedt RE (2006) How can we make food risk communication better: where are we and where are we going? J Risk Res 9(8):869–890

Lundgren RE, McMakin AH (2013) Ethical issues. In: Lundgren RE, McMakin AH (eds) Risk communication: a handbook for communicating environmental, safety, and health risks, 4th edn. Wiley & Sons Inc., New Jersey, pp 57–70

Renn O (2005) Risk perception and communication: lessons for the food and food packaging industry. Food Addit Contam 22(10):1061–1071

Rodriguez L (2007) The impact of risk communication on the acceptance of irradiated food. Sci Commun 28(4):476–500

Scholderer J, Frewer LJ (2003) The biotechnology communication paradox: experimental evidence and the need for a new strategy. J Consum Policy 26(2):125–157

Sperling D (2010) Food law, ethics, and food safety regulation: roles, justifications, and expected limits. J Agric Environ Ethics 23(3):267–278

Thompson PB (1997) Food biotechnology's challenge to cultural integrity and individual consent. Hastings Cent Rep 27(4):34–39

Verbeke W (2005) Agriculture and the food industry in the information age. Eur Rev Agric Econ 32(3):347–368

WHO (1995) Joint FAO/WHO expert consultation on the application of risk analysis to food standard issues. Geneva, March 1995

Chapter 10
Publication Ethics

Luis Adriano Oliveira

10.1 Introduction

Research is a driving force leading to social progress. Every good researcher has one basic objective in mind: to seek, seize, and use any possible opportunity to *improve*. In the particular case of industry – the food industry, for example – researchers struggle to produce higher-quality goods or provide better services with increasing efficiency, at a faster rate and lower cost, and minimizing the corresponding environmental footprint. All these goals mean improvement and the sky is the limit! In the context of such a permanent and demanding challenge, one question immediately arises: who can be a researcher nowadays? In other words, does one have to be a genius, like Leonardo da Vinci (1452–1519) or, more recently, Albert Einstein (1879–1955), to achieve a highly successful research career?

Presently, scientific knowledge no longer progresses essentially by revolutionary discoveries – as were the concepts of the catapult or helicopter (da Vinci), or the theory of relativity (Einstein) – but mainly through smaller steps, representing a variable degree of innovation. The most creative of these smaller steps may be seen as *disruptive innovation* (the laser beam, DNA, Higgs boson, etc.), while the majority of scientific progress is now mainly ensured by *incremental innovation* (small – although crucial – creative steps).

Nowadays most researchers are integrated in teams. Thanks to modern communication technologies, collaboration among research teams is almost free of charge and has virtually no geographical boundaries or time limitations. In these favorable

L.A. Oliveira (✉)
Mechanical Engineering Department, Faculty of Sciences and Technology of the University
of Coimbra, Rua Luís Reis Santos, 3030-788 Coimbra, Portugal
e-mail: luis.adriano@dem.uc.pt

© Springer International Publishing AG 2018

R. Costa, P. Pittia (eds.), *Food Ethics Education*, Integrating Food Science and
Engineering Knowledge Into the Food Chain 13,
https://doi.org/10.1007/978-3-319-64738-8_10

conditions, any averagely skilled person can be an excellent researcher. This is the good news. The bad news is a direct corollary of the good news. In fact, the number of researchers has increased exponentially in recent decades. Further, present researchers are distributed over a great number of scientific domains, very often with a high degree of specialization. Unfortunately, this dramatic growth of the scientific community has not been matched by a corresponding increase in its financial support: funding resources are necessarily limited.

Such an imbalance between the different rates of growth has led to direct and obvious consequences: (i) the pressure to publish is mounting up every day, as a good publication rate has become an essential condition for the survival of a researcher within the scientific community ("publish or perish", in the words of Harzing (2007)); (ii) competition among researchers to be published and/or funded has presently reached unprecedented levels; (iii) as a consequence of the huge number of manuscripts submitted, it is increasingly harder to have submitted articles accepted for publication in good, prestigious, high-impact journals.

In practice, the competition and pressure to publish has generated a scenario that is prone to temptation. Indeed, some researchers might be tempted to use quick fix strategies, or "shortcuts", in order to achieve a high publication rate rapidly. Such strategies often undermine the necessary respect for ethical principles, thus leading to questionable practices or even misconduct. Unfortunately, the growing number of cases, which have been both detected and reported, denouncing unethical behavior, demonstrates that this type of occurrence is much more frequent nowadays than one might initially imagine.

The present chapter addresses the ethical dimension of scientific publications. This question will be dealt with through different approaches, as illustrated throughout the various sections.

Firstly, a brief, general description of the research procedure is provided. Then, reference is made to the importance of publishing the research outcomes for the construction of scientific knowledge. The text continues by addressing several features linked to publications that frequently raise ethical concern: authorship, peer evaluation and the recognition of merit, the characteristics and values which support a good quality article, threats to those values, the role of error (ranging from simple, honest mistakes to misconduct involving deception), the consequences of unethical behavior, and strategies aimed at minimizing those consequences. Some concluding remarks are formulated at the end of the chapter, followed by the inclusion of bibliographic references.

10.2   The Research Procedure

Research is a driving force directly committed to promoting and developing social progress. Several definitions may be used to illustrate this concept. Following the one adopted in Oliveira (2012), "research is a creative and productive activity, based on ethical values and with the objective of achieving a profound understanding of relevant phenomena."

It is common practice to establish two categories of research, namely, *fundamental research* and *applied research*. Indeed, both are "applied", as they address needs and opportunities for improvement: fundamental research aims at being useful, sooner or later, although it is not necessarily intended to have a direct application in the short term; applied research, as its designation suggests, addresses a specific problem or question, hoping to find a concrete solution within a well-defined period of time.

Not surprisingly nowadays, where results are expected "the sooner the better", it is, thus, easier to get financial support for applied than for fundamental research. Either fundamental or applied, and regardless of the domain or subject at stake, what is research all about? In other words, what are the main steps any researcher must follow and take while conducting the research procedure?

First, she/he has to read the available literature devoted to the subject thoroughly, in order to understand it in depth. This *literature review* should make it possible to conceptualize the related phenomena and to formulate the corresponding societal context (how relevant is the subject to society?). Further, it should help identify the challenging questions, i.e., the problems related to the subject, for which a solution still has not been reported in the literature. Spotting these open questions is known as establishing the *state of the art* of the subject. Each open question identified by analyzing the state of the art is, in fact, an opportunity for the researcher to add his/her own contribution to scientific knowledge: among all the problems still lacking a solution, she/he must select *the question* that will be addressed – and hopefully solved – during his/her own research work.

This choice is particularly critical: indeed, it would be rather frustrating to realize, at the end of the research procedure, that one had simply been "reinventing the wheel", i.e., finding a solution for a problem that someone else has already solved. This is why the state of the art must be reviewed with great care and regularly updated during the research program. Once a still unresolved problem has been selected as the open question to address in his/her work, the researcher needs to reflect about it thoroughly, trying to come up with new ideas on how to tackle the problem and to find a way to formulate and implement a reliable solution (working hypothesis). As it implies creativity and innovation, this is perhaps the most challenging – and sometimes, to be honest, daunting – phase of the whole undertaking.

Formulating a strategy (generally designated as *methodology*) is followed by defining and creating (or selecting, if already available) the tools or *methods* necessary to implement it. Depending on the specific nature of the problem at stake, such tools may have different characteristics, ranging from theoretical methods – very often numerical simulations – to laboratory or field measuring instruments, or even questionnaires of different types.

It is then necessary to obtain *results* by applying the methods chosen to the problem. Results may be theoretical solutions, laboratory or field measurements, answers to questionnaires, etc. In order to gain confidence in the way the results were obtained, the first outcomes of the research methods must be tested, verified, and *validated*. This may be accomplished by comparing those outcomes with reliable references related to the same or to similar problems. Once the results have been validated, the conditions are met to systematically produce, explore, and interpret – i.e., *discuss* – the results that are relevant to address and resolve the main, initial open question.

Discussion of results naturally leads to the formulation of *conclusions*. Such conclusions may entirely or partly resolve the main question. In the latter – and more frequent – case, room is left for future work.

Finally, the conclusions obtained as a result of the whole research procedure contain innovation that has to be shared with the remaining scientific community and also with society in general. This is the role of the corresponding *publication*.

10.3   Publication: Why? Where? To Whom? What for?

Research outcomes can only be useful if they are shared with whoever may benefit from them. In order for the scientific community – and society as well – to acknowledge that they do, in fact, "exist", research outcomes need to be made public, i.e., to be *published*. If two researchers independently make the same important discovery, the entire merit will go to the one that publishes it first. This is why publication is important.

Writing and/or presenting strategies for a publication may vary widely, according to its market (Ashby, 2005). Depending upon the characteristics of the research and the corresponding outcomes – scope, goal, degree of innovation, potential readers or listeners, expected national or international impact, etc. – publications assume different formats and physical or virtual supports.

Typically, a publication includes the following sections (or chapters, if applicable): (i) title, (ii) abstract, (iii) introduction, (iv) methodology and methods, (v) results and discussion, (vi) conclusion, (vii) references, and (viii) other items, namely, keywords, table of contents, list of symbols, list of figures and tables, acknowledgments, dedication, appendices, etc.

10.3.1   *Master of Science (MSc) and Doctor of Philosophy (PhD) Theses*

Master of science (MSc) and doctor of philosophy (PhD) theses communicate the outcomes of the research work performed by postgraduate students, in order to obtain the corresponding academic degree. These documents are primarily addressed to the jury – supervisor(s) included – that will evaluate and rank the student's work. Generally, two versions are provided to the examiners, one printed on paper and the other in digital format. It is needless to stress the importance of such reports, as they are the first contact most elements of the jury have with the candidate, before the public defense of the thesis, or viva voce[1]. They are mainly intended to display: a demonstration of full mastery of the subject studied; awareness of possible alternatives to the one(s) used in the reported work; the main contribution of the student's work to the field of knowledge; or, in short, proof that the student deserves the award she/he wishes to obtain.

After final approval, MSc and PhD theses are accessible to a wider public, namely, through the archives of university libraries and alike.

10.3.2   Papers in Refereed Journals

The most efficient way to share and disseminate research outcomes consists of publishing them as an article in a refereed journal. *Journal* editors receive a manuscript that has been submitted for publication from the author(s). They then send the manuscript to – normally two or three – experts who are asked to give their opinion on how suitable the written report is in terms of acceptance for publication in the journal. The main criteria used by these experts – generally called *reviewers* or *referees* – to support their final recommendation to the editors are referred to in Sect. 10.4.1. They may advise that the manuscript be accepted for publication in its original form, accepted with optional or mandatory modifications, or rejected.

The *impact factor* (IF) of a scientific journal is referred to in Sect. 10.4.2. It is generally perceived as giving a good indication of its prestige and reputation within the corresponding field of knowledge. Aiming at maximizing visibility, all authors are interested in publishing their work in journals with the highest IF possible, within their scientific domain. Correspondingly, the degree of difficulty in having a paper accepted for publication in a scientific journal directly depends on the value of its IF.

Apart from referees, the public of an article in a journal includes other members of the scientific community and specialized elements of society. They use the published work to extract information relevant to their own work, or simply to increase their knowledge on that specific domain.

10.3.3   Conference Papers and Presentations. Posters

Thematic conferences are generally an excellent opportunity for members of the scientific community to meet and discuss not only completed but also ongoing research work. Ideas and suggestions are exchanged among participants, and authors are given a first impression of how likely it is that their work will be accepted for publication in a refereed journal, later on. The scientific merit of the manuscripts

---

[1] Further development on the viva voce can be found in Marshall and Green (2007) and Oliveira (2012).

submitted for presentation is evaluated by the conference's scientific committee, who may recommend acceptance for oral presentation with inclusion of a written version in the conference proceedings or, among other alternatives, suggest that the work be presented in the form of a *poster* displayed during the conference. In this latter case, one of the authors is supposed to be present and available to clarify any questions raised by conference participants.

The public and motivation for conference papers are essentially the same as those for articles in journals, although it is generally somewhat easier to have a contribution accepted at conferences.

10.3.4 Popular Articles and Other Means of Publication

When it comes to sharing research outcomes not only with scientists but also by addressing a wider range of people from society, who are not necessarily specialized in the area, popular articles are a very interesting option. Different media supports may be used to disseminate scientific knowledge by means of popular articles: the general or thematically driven press, television, radio, or magazines, among others. The market for such articles is frequently composed of readers or listeners that are interested in learning something about a new field or that are simply looking for entertainment. Accordingly, the style and contents of a popular article significantly differ from those of a refereed journal paper. In particular, reducing the use of specialized nomenclature to the strict minimum possible is a basic requirement.

Other means of dissemination of research outcomes include books, book chapters, research talks (seminars, workshop sessions, lectures, video conferences), or even – work-in-progress or final – reports on research projects.

10.4 Quality Assessment in Science Publication

Trust and confidence are keywords in the relationship among scientists and also between the scientific community and societal actors. In turn, high-quality research is the basic condition that ensures that research outcomes are worthy of our trust and confidence. In other words, scientific and social progress is directly linked to the guarantee that published research work is quality work.

Evaluation of quality in publications is therefore a requisite of paramount importance. There are essentially two stages where that assessment can be performed: (i) prior to publication and (ii) after publication. In the same order, the next two subsections are devoted to those phases of quality assessment.

10.4.1   Peer Evaluation of Proposals for Articles

When a manuscript reporting research outcomes in a scientific domain is submitted for publication in a journal, it is expected to contain innovation, compared to the corresponding state of the art. Journal editors are not necessarily specialized in all the scientific areas that are covered by their journal. In order to decide whether the manuscript deserves to be accepted for publication or not, they therefore ask advice from – generally two or three – invited experts.

Who is competent enough to assess the quality of a proposed paper that is situated at the limit of present scientific knowledge? Only peers of the authors of the manuscript are qualified to actually fulfill that requirement. This is the reason why quality assessment of proposed articles is usually done through the so-called procedure of *peer evaluation*.

Peer evaluation is also commonly used to assess the quality of proposals for scientific meetings (namely, national or international conferences) or to decide whether contractual research proposals are worthy of being financially supported.

The main role of experts in the peer evaluation of a manuscript consists of filtering irrelevant or bad quality proposals, if appropriate, and, prior to the paper's acceptance, correcting and/or improving some weak features of the submitted material.

Peer reviewers are members of the scientific community who have been invited to perform that high responsibility task, because they are considered to: be prestigious and acknowledged experts within the manuscript's scientific area; ensure absolute objectivity and impartiality in their judgment; and keep secrecy as long as necessary.

Generally, peer revision is focused on the following items: relevance of the proposed contribution within the scientific area; originality and innovation versus the corresponding state of the art; rigor in the way the reported research was conducted and reported; and quality and clarity of the manuscript (text, figures, tables, etc.).

Additionally, they are usually asked to express their degree of satisfaction, answering questions such as: Do the title and the conclusions adequately reflect the nature of the reported work? Does the manuscript contain superfluous or insufficient material? Are the conclusions sufficiently documented and supported?

Their final advice typically consists of recommending one of the three alternatives: (i) acceptance of the manuscript in the form it has been submitted; (ii) acceptance, if the authors perform a certain number of suggested modifications; and (iii) rejection. Naturally, the final decision belongs to those who have invited the reviewers' contribution.

10.4.2 Recognition of the Merit of a Published Work

The first and more obvious acknowledgment of merit in a published work is the corresponding authorship. Other common ways of expressing merit recognition are the section of acknowledgments and the list of references or even, if applicable, the dedication in the work.

Not every participant in a research work necessarily deserves to be included in the list of authors of the corresponding publication(s). Authorship implies fulfillment of at least one of the following conditions (cf. Elliot and Stern (1997); Stewart (2011)): conceiving and designing the research; conducting the research work; and writing the whole or part of the corresponding paper. In all cases, any author of a publication has to fully read it, approve it (in terms of contents and form), and take his/her part of responsibility for it. In short, authorship means contribution to the innovative and creative inputs of a publication. Routine work (as, for instance, maintenance and/or fine-tuning of measuring devices) does not justify, in itself, inclusion in a paper's list of authors: a note of thanks in the section of acknowledgments would be more appropriate to reward that type of contribution. The same applies to someone having financially supported the work.

The same principle is valid for the authorship of a published *patent*. Patents' authors are necessarily linked to the innovative concept of the patented work: as stated by Stewart (2011), using an interesting case study, the simple demonstration that the concept works in practice should not imply inclusion in the patent's list of authors.

However, authorship is not the only way to acknowledge merit in scientific publications. In fact, merit recognition can also be attributed by formulating the following question: How much visibility and influence has the published work within the scientific community and, ultimately, within society in general? In other words: What has been the real *impact* of the publication within its potential public? If I write an article where I make a reference to another publication, this means that I have read the publication, and I found it interesting enough to cite it, i.e., the publication had impact upon my own work.

It would be quite appropriate to adopt a perfect, fair, and universally accepted procedure to weigh the impact of scientific publications. Unfortunately, such an ideal criterion does not exist. Instead, several criteria have been suggested, aimed at quantifying the impact of publications, each one with its own advantages and shortcomings. This is what is generally named *bibliometry*.

The impact of a scientific journal directly results from the impact of its published papers. The commonly accepted measure for the impact of a journal is its *impact factor* (IF). This numerical index quantifies the annual average number of citations of the articles that are published in the journal (see also Garfield, 2006).

Merit recognition of individual scientists is generally attributed by taking into consideration the following factors: the IF of the journals where their papers are published; the number of published papers that they have authored or co-authored; and the *number of citations* that their publications have received in other publications.

Conference papers or communications are also included in this form of individual evaluation.

Named after its creator, Hirsch (2005), the *h index* is probably the most well-known indicator, aiming at measuring both individual scientific productivity and citation impact. More precisely, the h index of an individual researcher will be h = n if she/he has authored or co-authored N publications, n of which have (individually) been cited at least n times, the remaining (N-n) publications not having been (each of which) cited more than n times. For example, if an author has authored or co-authored 20 articles, 8 of which have (individually) received at least 8 citations, then his/her h index is 8.

The calculation procedure of the h index shows that it cannot decrease along time, therefore favoring the more senior researchers. On the other hand, h index comparisons may only be made among researchers within similar scientific areas, as the number of potential readers, and thus citations, may vary significantly from area to area.

Aiming to overcome the unfair relative advantage of senior researchers, alternatives to the h index have been proposed, namely, the *m index*. This index is calculated by dividing the h index by the number of years since the first author's publication. Other indices (b, h-b, c, s, etc.) are available, and many others will certainly be proposed in the future.

Internet databases, as ISI Web of Knowledge® or Elsevier Scopus® are available that display the h index of authors having been published in journals indexed in Journal Citation Reports (JCR®), which is an annual publication by Thompson Reuters. The program Publish or Perish®, authored by Anne-Wil Harzing (Harzing, 2007), is available from the database Google Scholar®: it provides free access to cited articles, as well as a regularly updated list of the most common metric indices for individual authors.

Metric culture for quality assessment in science publication has obvious limitations and has therefore been widely criticized (cf. Lawrence (2003); Parnas (2007); Amin and Mabe (2000); Zare (2012), among others). In fact, the most used indices do not necessarily: take into account the number of papers' authors; distinguish citations from auto-citations; consider the context of citations, which may well be critical or simply used to signal the existence of errors; or detect "mutually agreed citations" (I cite your paper, you cite mine...).

In short, metric is a useful first indication to evaluate the productivity and impact strength of individual scientific researchers. However, that criterion should always be supplemented by a thorough, qualitative analysis of their scientific profile and published footprint.

10.5 Values Underpinning Publication Ethics

What are the main characteristics that make the difference between a good and a bad publication? A high-quality publication has to be socially relevant, well written, original, and innovative. No question about that! However, that is not the whole picture: in parallel with these qualities, a report of scientific research only deserves to be published if it is *ethically irreproachable*. This involves a wide range of questions and related ethical values, as discussed in the present and the following sections.

The universal dimension and importance of values underpinning publication ethics is well illustrated by the outcomes of three international conferences that were recently organized on research integrity: Lisbon, Portugal (2007); Singapore (2010); and Montreal, Canada (2013). The corresponding final declarations (Mayer and Steneck (2007, 2010); Anderson and Kleinert (2013)) are relevant documents, available on the subject.

10.5.1 Reproducibility

Let me make an important point by using the analogy of a recipe. Here is the recipe for cold melon soup with bacon.

Ingredients: diced bacon; one melon; cream (250 ml); salt to taste; sugar to taste.

Preparation: mash the melon; add the cream, salt, and sugar; stir it well; filter through a sieve and place it in the refrigerator; fry the bacon lightly; serve well chilled; add the bits of bacon.

Now try it, the result will be a disaster; the taste is simply sickening. Why? Something is wrong. In fact, the recipe should read as follows:

Ingredients: diced bacon; one melon; *one lemon*; cream (250 ml); salt to taste; and sugar to taste.

Preparation: mash the melon; *squeeze the lemon into the mixture*, add the cream, salt, and sugar; stir it well; filter through a sieve and place it in the refrigerator; fry the bacon lightly; serve well chilled; add the bits of bacon.

The point is that lemon was missing in the first recipe. It may seem a small detail, but, in practice, it does make a huge difference! Now the taste is perfect. This is about reproducibility. Anyone trying to reproduce the original cold melon soup using the first recipe would never achieve it! Researchers very often use other people's publications as a starting point for their own work. If relevant "details" are wrongly reported or simply missing from a published work, newcomers may spend months, or even years, trying to reproduce something that is simply non-reproducible! Deliberate or not, this lack of ethics can have and, in many cases has had, very serious consequences in terms of time, money, human resources, and general confidence in science.

Ethical scientific research means ensuring a full respect for values in science, during the whole research procedure and also when it comes to publicly disseminating, i.e., publishing, the research procedure and the corresponding outcomes. Together with reproducibility, which has already been stated as crucial, other values must be mentioned that underpin the respect for ethics in a scientific publication.

10.5.2 Verifiability

Verifiability is naturally linked to, but should not be confused with, reproducibility. All the information underlying relevant conclusions in a publication should be carefully organized and stored for a reasonable period of time (no less than about 5 years). This allows, whenever necessary, any interested person to check and verify the way those conclusions were drawn. In fact, sometimes it is not possible – or even very relevant – to include every small detail of all the arguments that legitimize all the conclusions of a publication. But that information is relevant and should be easily accessible in case of need. This is also a defense against possible suspicions of fabrication and/or falsification for the authors, as we shall see later.

10.5.3 Quality, Dependability, and Responsibility

Quality and reliable work means that all those participating in the research procedure (including researchers, technicians, and elements of staff) give their best to both the execution and reporting of the research. Full dedication is absolutely essential to ensure confidence in the research outcomes published. Accordingly, authors must take full responsibility for what has been published with their name on it.

10.5.4 Clarity

It is generally acknowledged, by both scientists and society, that publications are the most efficient way of disseminating potential advances in science. The role of any publication is, thus, to convey a message. That message only manages to get through if it is clearly formulated. A nebulous message is absolutely pointless: instead of making its point, it is confusing and generates suspicion. Although a clear and well-presented publication does not necessarily ensure the quality of the research contained therein, a high-quality research can be entirely spoiled if conveyed in a messy text! Quoting Boileau-Despreaux (1674), "If you clearly understand something, you can explain it clearly."[2] Clarity inspires confidence.

---

[2] In French: "Ce qui se conçoit bien s'énonce clairement."

10.5.5   Honesty and Transparency

A scientific publication is meant to report some research work accurately, as well as the corresponding outcomes. The authors of an article are the first to know how confident they are in their results and also the degree of innovation that is actually contained in those outcomes. Proclaiming the merits and assuming the limitations of the work published must be the permanent exercise in transparency and honesty, never overestimating the former or underestimating the latter.

10.5.6   Neutrality

Scientific knowledge is inherently objective and universal. There is no room in science for bias due to undisclosed personal beliefs or interests. In other words, neutrality is an essential value in the never-ending search for scientific knowledge. However, this does not preclude the positive influence of subjective factors like local culture, or even religious faith. Returning to the analogy related to food, cuisine is an excellent example of how local culture and tradition may be enriching. Integrity of, say, culture-driven research and its publication is by no means affected if the influence of the culture is clearly disclosed. Neutrality is also supposed to be maintained when it comes to evaluating the merit of a manuscript, in order to decide whether it deserves publication or not.

10.5.7   Criticism and Openness to Debate

Quoting an African proverb, "If you want to go fast, go alone; if you want to go far, go together." Scientific research is a perfect illustration of the wisdom of this ancient thought. In fact, criticism and openness to debate are key values in the permanent search for improvement. In the absence of those values, man would probably still think that the Earth was the center of the Universe! Openness to discussion and cooperation are achieved by what we now call teamwork. Published works frequently result from international cooperation, involving coauthors from different regions of the world.

10.5.8   Acknowledgment of Other Peoples' Work

Cooperation can only be efficient if everyone is willing to acknowledge the work of others, should they be scientists or not. This acknowledgment may, of course, take different forms. In a publication, authorship, bibliography and the acknowledgment section are the three alternatives most frequently used for that purpose.

In short, values like reproducibility, verifiability, quality, dependability, responsibility, clarity, transparency, honesty, criticism, openness to debate, and acknowledgment of other peoples' work are absolutely fundamental to ensure integrity when doing and disseminating scientific research. Other items, such as scientific humility (we know a lot, but are ignorant of so much more) could be added. However, the list would be never-ending.

10.6 Publication: Threats to Values

In real practice, the values underpinning publication ethics may be actually challenged – or even put at risk – for several, different reasons. Analyzing some of them will provide a better understanding of the importance of such values. This is the main goal of the following paragraphs.

10.6.1 Pressure to Publish

Pressure to publish is a direct consequence of the present imbalance between the growing number of scientific researchers and the limited resources that are available to financially support their work (Sect. 10.1). Only those researchers or research teams, who manage to have their work published, are sponsored by public or private institutions. On the other hand, only those receiving financial support can do research worthy of being published. As a condition to being accepted and acknowledged within the scientific community, researchers and research teams thus desperately need to publish. The dramatic nature of this "fight for survival" is well illustrated by Harzing's (2007) well-known expression "publish or perish."

In practice, a great number of proposals for articles of varying scientific quality are submitted for publication every day. Publishing agents do not have enough room for, or interest in, all these proposals. A filtration system is thus badly needed. As a result, it has become increasingly difficult to succeed in having submitted manuscripts accepted for publication in high-impact journals. The selection procedure has been previously mentioned in this chapter (Sect. 10.4).

Therefore, fierce competition is firmly installed among scientists and scientific teams. While this may be a positive factor to ensure quality, some participants may be tempted to achieve success more easily by using "shortcuts", undermining or simply ignoring the ethical principles underlying the integrity of science. Consequences of this unethical behavior may be unpredictably serious, as referred to below.

10.6.2 Conflicts of Interest

Each time personal interests interfere with professional obligations, there is a conflict of interest. Conflicts of interest may jeopardize the necessary objectivity of research publication, biasing the work description, the publishing of the corresponding outcomes, or both. Mostly centered on examples from the university environment, the following situations illustrate potential permeability to that type of conflict.

Consulting Activity It is difficult for a researcher to ensure impartiality if requested to publish an article where she/he assesses the quality of a product made by the same company that sponsors his/her research. The same applies if a researcher agrees to be the expert that signs a report analyzing the causes of an accident involving an entity directly linked to his/her work.

Financial Interest Doing and publishing applied research concerning a product that is, or will be, manufactured by a company where the researcher, or his/her close family, has a direct financial interest also involves a conflict of interest: even inadvertently, the unbiased scientific objectivity necessary may be at risk.

Peer Evaluation Delicate situations arise, for instance, when a reviewer of a manuscript submitted for publication in a journal realizes that the manuscript contains more innovation than the reviewer's own results or when a reviewer is evaluating the proposal of an article that is authored by one or more of his/her close colleagues.

Openness to Publication Researchers and institutions sponsoring research generally have different interests in making results public. In fact, the financing institution may wish to keep the result outcomes secret, at least for a certain period of time, in order to avoid competition. On the other hand, researchers are interested in publishing their results as soon as possible, therefore valorizing their *curriculum vitae* (CV). The conflict of interests here is clearly apparent. The same is valid for sensitive research (classified material) involving industry, political strategy, or military defense. Copyrights and patents also illustrate how intellectual property can be a source of conflicts of interest.

Researcher's Social Responsibility The researcher's interest in contributing to scientific knowledge may end up seriously harming society and/or the environment. Nobel's discovery of dynamite was initially intended to be useful in civil and mining engineering. Nowadays dynamite is still used with great success, namely, in mineral extraction and the construction of tunnels. However, history also clearly shows it has been used as an effective killing device. The same applies to research centered on chemistry, biology, lasers, or nuclear energy: the progress of science and its useful applications (e.g., in medicine and in communication technologies) are in clear conflict with the use of that knowledge for the fabrication of weapons of mass destruction.

Research Involving Human, or Animal, Subjects The researcher's willingness to obtain conclusive results rapidly and worthy of publication (medical research and clinical trials are good examples) may collide with the interests of the human (or animal) participants. Understandably, the participants are mainly concerned about their security and comfort, dignity, privacy, and the right to vet prior to publication and/or to remain anonymous.

The problem is even more delicate when participants are particularly vulnerable: this is the case of children, elderly people, or pregnant women. However, it is also true for animals, as they are unable to defend themselves and their rights.

10.6.3 Conflicts of Commitment

Conflicts of interest are often linked to conflicts of commitment. In this case, the central issue is about managing one's personal agenda, in terms of time, interest, and effort. If I say yes to everything, accepting any invitation to perform different tasks, something will end up by being left behind. The reason is quite simple: each day only has 24 hours. Conflicts of commitment are often patent through physical absence from the workplace. Some examples will follow, again mainly centered on the university environment.

Accepting Engagements in Different Institutions In this context, senior elements are more prone to conflicts of commitment: owing to their knowledge and experience, they are more frequently invited and requested than junior ones. Consulting activities in other institutions are generally compatible with research and even teaching. However, other tasks like participation in a company's management or administrative board are very demanding, both in time and dedication, and can thus be a source of concern.

Progress in the Research Career Many researchers also have teaching and administrative obligations. In the particular case of universities, a good publishing record highly valorizes a CV, in terms of career progression. Priority is thus attributed to publication, sometimes underestimating other interests, namely, those of the students who also need to benefit from their teacher's or supervisor's attention.

10.6.4 Insufficient Culture of Ethics

If ethical values are not thoroughly promoted, a growing sense of impunity is most likely to occur, directly affecting the quality of both research and the way it is reported. This may be due to a lack of *teaching ethics*: researchers, in particular the youngest, may have not been sufficiently warned of the dangers and consequences of breaching their ethical obligations. But it can also be attributed to the absence or vagueness of *normative texts* providing principles and guidelines for the ethical development and dissemination of scientific research. Doubts concerning what is acceptable and recommended and what is not can then be easily installed. Finally, if *discipline* is not at the basis of the work leading to publication, more or less chaotic situations may occur, where "shortcuts" are frequently needed to overcome unexpected difficulties that are permanently arising and accumulating. Such quick improvisations generally result in undermining or simply ignoring the basic ethical requirements that ensure the integrity of science.

Some features related to publishing scientific research are particularly sensitive, as they raise questions that are directly linked to the issue of ethical values in science. The role of error in conducting and/or disseminating science is an important example of those issues. It is the central focus of the following section.

10.7   Error in Scientific Publication

Ethical practice in scientific research and publication is a permanent quest for perfection, aiming at maximizing benefits and minimizing shortcomings. Unfortunately, ethical principles are not always fully respected, and even when they are, imperfection often remains unavoidable. Some of the most frequent errors occurring in scientific publications are mentioned next.

10.7.1   Honest Mistakes

Human beings are inherently imperfect. Scientific knowledge is and will always be limited, simply as a result of human nature. The tools (methods) used in scientific research are also conceived and implemented by humans. The performance of those tools (robustness, rate of convergence and precision of numerical methods, precision and power of resolution of measuring methods, and the confidence interval of statistical approaches) is progressing every day. In short, scientific truth is not an absolute concept. Instead, it evolves as the research capability continuously evolves.

Even if full respect of ethics is ensured, some of today's published results may need to be corrected sooner or later. For example, the most elementary particle of matter has successively been claimed to be the atom, the electron, and the Higgs boson, just to mention a few of the successive, intermediate statements. Were those intermediate claims correct? No! Should their authors be blamed for them? Absolutely not! Why? Simply because they had done their best, basing their contribution on the (at that time) state of the art, using the best research tools available at that moment. No ethical precept was infringed during their research.

This type of error is generally called an *honest mistake*. The author's prestige and ethical reputation remains intact. However, this will only happen if one condition is fulfilled: that it has to be acknowledged and corrected (see Sect. 10.9.2).

By definition, an honest mistake is unintentional. It in no way affects the integrity of scientific knowledge or the corresponding publication (see also NAS, 1995).

10.7.2   *Errors Due to Misconduct in Research Not Involving Deception*

Some mistakes are unintentionally committed in research, leading to erroneous publications, which could – and should – have been avoided. Although not deliberate, such errors may have very serious consequences, as will be pointed out later. They may be due to various causes, some of which are referred to in the next paragraphs.

Negligent Work Haste, carelessness, and inattention are typical factors that may lead to this kind of errors. In a way this is equivalent to underestimating the so-called Murphy's law, which states: "if something can go wrong, it will." Lack of discipline and conflicts of commitment (Sect. 10.6) are often at the origin of negligent work.

Incompetence This happens when research and publication tasks are performed by people that lack the appropriate or sufficient, academic experience or background to perform them. Again, this is an underestimation of the wisdom contained in Murphy's law.

An example is that the real authors of a research work do not feel they have sufficient skills, patience, or time to produce the corresponding publication. They then decide to give a "ghost writer" the task of writing the text. Someone not competent in the scientific area covered by the article may inadvertently distort the message to be conveyed. Obviously, the necessary dependability and responsibility are far from ensured by such behavior.

Other Errors Resulting from Questionable Practices Although not involving the intention to deceive, other forms of questionable scientific conduct can originate errors of varying gravity. The list is virtually endless: inadequate management of the article's authorship, injustice in the acknowledgment of scientific merit, accumulation of delays due to a lack of discipline, and other forms of mismanagement and insufficient management of the scientific obligations involved are some examples worthy of mention, among many others.

10.7.3   *Errors Due to Misconduct in Research Involving Deception*

Errors generated by deliberate misconduct are aggravated by one additional factor: they are committed with the intention to deceive. Scientific integrity is then confronted with a threat of the utmost gravity. Misconduct in science involving deception may be divided into three basic categories: fabrication, falsification, and plagiarism (FFP).

Fabrication This is probably the most serious and harmful way to cheat in science. Fabrication consists in inventing data and/or results that do not exist, and then publishing them, claiming they are the real outcomes of research work. Although not totally unprecedented, fabrication rarely involves the whole publication from A to Z: a great deal of imagination, experience, and even cynicism would be required for that. Instead, what happens most frequently is that small gaps are filled with invented material.

For instance, imagine I needed to perform a number of experiments to confirm some results I was expecting – and perhaps almost sure – of obtaining. For some reason (lack of time, insufficient financial support, defective experimental setup, etc.), I was not able to perform those experiments. Even so, I publish the "expected" fabricated results stating that they were actually obtained from my experiments. In this example two lies are involved: fake experiments and fake results!

Falsification Results obtained from research can be divided into two main categories: favorable and unfavorable. *Favorable results* tend to confirm the researcher's initial suppositions, thus supporting a "clean" publication that inspires confidence and leaves little room for doubt. The opposite happens with so-called *unfavorable results*. Journal editors are generally more willing to publish favorable than unfavorable results (unlike in politics, favorable news tends to sell more than unfavorable in science!).

In the rush to publish, the temptation is there: by means of "small" modifications, unfavorable results can be turned into favorable ones, thus increasing the probability of the manuscript being accepted for publication. This is falsification. Quoting the definition used by Oliveira (2013), "falsification is about modifying the researcher's or others' data or results, thus leading to interpretations and conclusions different from those which would have been drawn from the original, unmodified material." Suppressing or omitting relevant information is also falsification.

Unfortunately, modern software makes it easy to falsify data or results. It is indeed not difficult to: suppress or modify the position of one or more points on a curve; replace an experimental curve by another, smoother one, generated by an equivalent mathematical expression; or artificially reduce experimental dispersion by modifying the horizontal and/or vertical scale of a figure without previous notification. Apart from being dishonest, these two forms of misconduct can – and frequently do! – mean a lost opportunity to produce good science: most of the relevant and creative steps forward, in scientific knowledge, come from unexpected results (or "anomalies"), not from expected ones. Just consider the example of Marie Curie (1867–1934): had she simply suppressed the unexpected frequencies of polonium and radium from her radioactivity spectra, those two elements would probably still be undiscovered nowadays!

The wisdom of organizing and storing relevant material supporting research conclusions, highlighted in Sect. 10.5.2, is now more apparent: in addition to the obvious advantages, it is a wise way to protect against possible suspicions of fabrication or falsification.

Plagiarism This class of misconduct consists of intentionally reproducing ideas, texts, data, figures, expressions, or material similar to that which was originally authored by someone else and then failing to identify the corresponding source and thus omitting to give the credit due. In other words, I am plagiarizing if I report or publish the work of someone else as if it were my own work. Plagiarism is thus about stealing authorship.

The Internet is presently an important source of temptation to commit plagiarism. Indeed, by inserting the appropriate keyword(s) in an Internet search engine, an almost infinite – although not necessarily reliable! – world of material for further consultation about any possible subject is nowadays immediately available and free of charge. Together with falsification, plagiarism is perhaps the most common form of intentional misconduct affecting publications nowadays. Many cases have been detected recently, suggesting that probably many more remain undisclosed. Similar to fabrication and falsification, plagiarism rarely involves the whole extent of the work published. Instead, it has been frequently spotted in small parts, namely, sections, paragraphs, figures, or simple sentences.

Taking the example of a sentence, one question immediately arises: How many words from the original text can be reproduced, in the same order, before the text is considered to have been plagiarized? Two? Three? Four? Stewart (2011) wisely replies that what really matters in a newly published work is not the number of words that may have been (sometimes inadvertently) reproduced but their quality and originality. However, if a "considerable" number of words are actually intentionally reproduced, any possible ambiguity is clarified by simply putting those words in quotation marks.

Another way of committing plagiarism is to plagiarize one's own work: this is called *self-plagiarism*. The most striking scenario consists of publishing exactly the same article in, say, two different journals. But some more subtle circumstances should also be considered. For example, if one author publishes the same scientific contribution in two distinct articles, this is also self-plagiarism. In fact, the crucial point is that any published work must be original. If part of a previously published work is published again in another article, the second publication is no longer original: plagiarism is therefore committed (self-plagiarism, if involving the same author(s)).

Other Errors Resulting from Misconduct in Science In the present context, the distinction between questionable practices and misconduct is based on intention: questionable practices are essentially unintentional, while misconduct is deliberate and directly involves the intention to deceive. The consequences of the latter are thus much more serious.

Again, the list is virtually endless. Some examples of misconduct in science are obstructing others' work (namely, due to envy or jealousy of their success), fabricating false accusations, exploiting others' work for one's personal benefit, ignoring commitments (breaking the pact of secrecy or deliberately not respecting deadlines), or ignoring established procedures (for instance, convening the media to publicly disseminate research outcomes that have not been previously validated within the scientific community).

Finally, other forms of misconduct may occur that directly imply recourse to the legal system, namely, misappropriation of funds, persecution or reprisal of whistle-blowers, personal harassment, or nepotism. This type of criminal behavior obviously falls outside the scope of the present chapter.

10.8 Consequences of Unethical Practices

Threats to values in science (the main subject of Sect. 10.6) may materialize in the form of unethical behavior, originating all types of consequences, some of them particularly serious, and even dramatic.

10.8.1 The Spreading Effect

Typically, misconduct in research originates at the individual level. If not properly dealt with, the corresponding effect tends to spread to successively larger scales, namely, the researcher's team, the scientific community, and, ultimately, society as a whole. Thus, the spread of consequences can be represented schematically in this chain: researcher –> team –> scientific community –> society.

10.8.2 Loss of Confidence

Society needs to trust science in order to profit from scientific knowledge and to accept the duty of supporting scientific research. It is therefore of paramount importance to the scientific community that it preserves the integrity of the scientific knowledge it produces, thus showing that it is deserving of confidence, both internally and externally. If confidence is lost or seriously eroded, the whole relationship between scientists and society is put at risk. Thanks to its own detection and self-regulating mechanisms (some of which are referred to in Sect. 10.9), the scientific community has a considerable capacity for dealing with errors within its boundaries, therefore preventing damaging effects from spreading to society via media supports like television, radio, or journals. However, certain forms of misbehavior are just too serious for their consequences to be kept "in-house."

10.8.3 Questionable Practices

Questionable Practices or unintentional errors (due to negligence, incompetence, or suchlike), as well as errors resulting from questionable practices (see Sect. 10.7.2), generally do not affect the integrity of science. They therefore tend to be dealt with by science's self-regulating mechanisms. Nevertheless, some of those errors may have huge, or even life-threatening, consequences (one example: the deterioration of food due to hasty research leading to an incorrect storage procedure). In this latter case, the spread of the consequences to society is obviously unavoidable. Although insurance policies may cover part or all the material responsibility of those who commit such type of errors, they will never cover the consequences of their lack of ethical responsibility.

10.8.4   Misconduct

In contrast with unintentional, unethical practices, errors resulting from fabrication, falsification or plagiarism (FFP) all have the direct intention to deceive in common. In contrast with inadvertent mistakes, FFP errors usually affect – often critically – the integrity of science. They are almost impossible to contain within the scientific community and therefore easily spread to society through the media, often in the form of scandals. Confidence in scientific knowledge is betrayed, and the public will to financially support it is also put into question.

Apart from practical consequences (remembering the example from the food industry, fabrication and/or falsification of research outcomes may be lethal and affect a whole population), FFP practices severely damage the whole procedure of quality assessment in research. Namely, artificial manipulation may lead to the publication of bad quality – or even dangerously wrong – papers and to their acceptance as "science." Further, those fraudulent papers may be taking the place of good quality ones that, in comparison, have been rated with lower marks and therefore rejected. By stealing the authorship of others' work, plagiarism gravely affects the whole system of the recognition of merit (and attribution of financial support) in scientific research and publication. Other errors, also involving deception and resulting from misconduct in science, were mentioned in Sect. 10.7.3. The consequences are generally critical and easy to predict. For example, by exploiting a PhD student's work for his/her personal benefit, a supervisor can prevent the student from focusing on the main research subject of the thesis, therefore critically damaging the student's work and career. A comparable effect results from ignoring or abandoning a supervisor's obligations (often as a consequence of conflicts of commitment) or, even worse, from the misappropriation of the authorship of a student's work.

In contractual research projects involving the partnership between research institutions and industrial companies, breaching the rules and ignoring commitments (namely, disregarding the contract of secrecy) may seriously damage the expected benefits of the research outcomes. Mutual trust is thus irreversibly lost. Unfortunately, recourse to the legal system is then often unavoidable.

10.8.5   *The Assessment of Quality*

Understandably, whether intentional or not, inadequate or unfair assessment of quality in scientific publication can lead to different levels of human resentment. However, although this, in itself, may have a very pernicious effect, errors in such a crucial phase of scientific production may also seriously damage the integrity of science, as well as social well-being or even safety.

Any unethical practice occurring during the peer-review phase may create profound injustices within the whole system of quality assessment concerning the manuscripts submitted for publication. Again, bad quality papers may be accepted, and good ones rejected. Causes for faulty reviews may range from a simple lack of motivation (not enough time devoted to ensuring a thorough analysis and evaluation) to the deliberate intention of favoring some authors and/or mistreating others (e.g., by misappropriating the submitted work for the reviewer's personal benefit). Excessive feelings of competition, jealousy, or revenge may also originate biased and unfair peer-review work. Such circumstances clearly illustrate how conflicts of interest and conflicts of commitment can gravely harm the integrity of scientific knowledge in its entirety.

Public interest and safety may also be at risk when transparency and impartiality are threatened by other manifestations of conflicts of interest. If we consider an example taken from the field of consultancy, it is not difficult to conclude that there is a natural tendency to publish results emphasizing favorable outcomes while hiding any unfavorable ones. This is particularly clear if the manufacturer is paying the consultant for his/her evaluation! Consumers might thus be exposed, without sufficient warning, to side effects that are potentially dangerous for public health.

10.8.6   *Social Responsibility*

Research outcomes with a wide range of potential applications can be equally problematic. In fact, the misuse of some of those discoveries may have disastrous consequences: together with the cases of dynamite and lasers, discoveries in areas such as chemistry, biology, and nuclear energy were already mentioned in Sect. 10.6.2. That same section also highlighted the potential costs of misbehavior in research involving human or animal subjects.

Are all these consequences of misconduct (and so many others that could have been mentioned at this stage) unavoidable and inescapable? Not necessarily: some causes of error can be anticipated and even avoided. However, when this is not feasible, it is still often possible and realistic to minimize the corresponding consequences. This positive message is the main focus of the next section.

10.9   Avoiding or Minimizing Negative Impacts

The ethical values underpinning the integrity of both the research procedure and the publication of its outcomes are continuously challenged by different kinds of ever-evolving threats. On the other hand, the scientific community has a remarkable ability to learn from its own errors, thereby constantly bolstering its self-regulating mechanisms. The good news is that, globally, the pace of the spread, refinement, and diversification of all kinds of possible threats is being mitigated by our ability to deal with them. Indeed, several realistic and efficient strategies are available to those who produce, publish, and assess scientific knowledge in order to avoid, or at least minimize, the negative impact of questionable practices or misconduct in science. Such strategies are the main focus of the following paragraphs.

10.9.1   Managing Conflicts

Conflicts of interest and conflicts of commitment are often mutually linked. The former are generally related to activities involving money, while the latter relate to the management of time and personal agenda. Although they cannot be always completely avoided, it is still possible to control them, thus minimizing the corresponding consequences. Transparency, establishing a scale of priorities where loyalty to the main employer stands in the first place as the main criterion, and, above all, the ability to say no are the key attitudes to dealing with such situations of conflicting interests.

Ethical Disclaimer If a researcher accepts publishing an article assessing the quality of an industrial product, it is, nowadays, common practice to request that the author provides an ethical disclaimer. The disclaimer is an announcement where the author displays his/her recent CV activities, thus making it possible to check if there has been any recent relationship with the product's manufacturer. If applicable, transparent information is also required about who pays, and for what, in the whole assessment procedure. The same logic applies to consulting activity.

Research funding institutions may detect – and then denounce – the occurrence of a conflict of commitment simply by cross-checking data provided by a researcher concerning the percentage of personal working time allocated to different projects or activities she/he is involved in: naturally, the total amount cannot exceed 100%! When confronted with a potential conflict, three main positive reactions are available: (i) never ignore or conceal the problem, instead display it to those concerned; (ii) whenever possible, avoid the problem circumventing foreseeable incompatibilities; (iii) if that is not feasible, just say no, desist, thus eliminating any source of ethical discomfort.

The frontier between what might or might not imply a situation of conflict is not always obvious and clear to someone confronted with an invitation to perform a certain task. Stewart (2011) wisely suggests what he calls the "newspaper rule": if the situation involving me were displayed on the front page of my local newspaper, would I feel comfortable or embarrassed? In any case, conscience is – always! – the best and ultimate referee.

Conflict of interest statements are presently on hand to help manage the risk of conflicts in all sorts of institutions, situations, and scientific areas. For example, the interests of participants in partnerships (namely, within industry or between industrial companies and research institutions) are safeguarded by protocols of cooperation. When appropriate, *non-disclosure agreements (NDAs)* ensure a confidential relationship between the parties, during the negotiation phase or throughout the duration of common research projects or other forms of cooperation or trading, therefore protecting confidentiality and proprietary information. If they are appropriately drawn up[3] and fully respected, such regulatory texts can minimize or even eliminate what otherwise would be potentially conflicting situations. This obviously has direct implications concerning the commonly agreed publication policy.

Ethical statements are also of paramount importance to provide guidelines for research and publication involving human (or animal) subjects. Historically, the Nuremberg Code (1949) and the Declaration of Helsinki (1964) are acknowledged as two pioneering texts containing the principles of research ethics for human experimentation and for medical research involving human subjects, respectively. Since then, regulatory texts and codes have been constantly improved and adapted to all kinds of specific, practical situations. The logic of these protocols aims at protecting the interests of the target population. For the particular case of clinical trials, that safeguard typically implies: (i) the participants' formal, written authorization; (ii) risk minimization through suitable, previous preparation (including the analysis of similar experimental results); (iii) thorough and transparent calculation of the risk/benefit relation; (iv) insightful selection of the target population (group of participants must be statistically representative and must not include vulnerable elements); (v) participants must be thoroughly informed (about possible risks, advantages, disadvantages, etc.); (vi) the trial has to be monitored in real time, with the possibility of being suspended at any moment; (vii) aid must be permanently available; (viii) the dissemination of results must respect dignity, privacy, and the right to vet prior to publication and/or to remain anonymous. For clinical trials involving the participation of animals, the focus is placed on minimizing suffering and the risk of death.

Concerning the *social responsibility of researchers* in scientific innovation, it is their ethical duty to carefully evaluate and ponder the possible consequences of their research outcomes. This may lead to self-imposed restrictions when publicly reporting their work. In my personal opinion, the existence of potential risks should

---

[3]The preparation of such sensitive and complex texts is generally done with the aid of lawyers, who are also asked to subsequently ensure that the partners abide by the commonly accepted agreement.

not be a reason for them to stop innovating (nuclear medicine would not be presently available if research on nuclear energy had been stopped because of the fear of its application in atomic weapons). Instead, decision makers should be promptly and thoroughly warned about any potential risks. In fact, most of the time public threats are not a direct result of innovation: they are a consequence of the misuse of innovative outcomes.

In a *research team*, the early definition and negotiation of who does what and when (for instance, the list of authors of the foreseeable deliverables to be published) can promote general efficiency and avoid potential human resentment. A clear and open dialog among the researchers involved is indispensable and should always be fostered by the team manager.

Peer Review Until recently, it has generally been accepted that reviewers of manuscripts submitted for publication have more "right" to remain anonymous than the papers' authors. Some reviewers have actually abused the feeling of immunity resulting from this privilege, resulting in cases of insufficient time and care being devoted to their review, and lack of courtesy in their comments to the authors. In order to safeguard interests of both reviewers and authors, some editors are presently questioning the legitimacy of that policy, thereby requesting full identification or full anonymity for all actors (authors and reviewers) during the review process. The second option, known as a *double-blind peer review*, has the merit of avoiding mutual favors or reprisals, or even the subsequent retaliation for negative assessments. Increasing transparency in the whole process has been promoted by the right authors often now have to full access of the reviewers' comments and final decision, so that, if necessary, they are able to refute them. In short, although far from being perfect, peer review is subject to ever-increasing improvement and scrutiny: it is still the best available form of assessing the quality of scientific research and publication.

10.9.2 Dealing with Errors in Scientific Publications

Honest Mistakes The integrity of science is by no means put at stake by the possible occurrence of honest mistakes in scientific publications (Sect. 10.7.1). The same applies to the ethical and scientific reputation of those authors that are responsible for such type of unintentional, erroneous statements. Nevertheless, this depends on one condition: as soon as the error is detected, it has to be acknowledged and corrected immediately, preferably in the same support (e.g., journal) that was used to publish the wrong conclusions.

Errors Involving Misconduct On the other hand, intentional errors like FFP (Sect. 10.7.3) do affect the integrity of scientific knowledge and of those responsible for them. Peer reviewers are not specifically expected to detect intentional errors in the manuscripts they are examining. However, if for some reason they do, or simply spot any reason for suspicion, it is their ethical obligation to immediately alert the journal editor(s). Fabrication and falsification often come to light because of the non-reproducibility of the research published.

Software for the detection of plagiarism, as well as self-plagiarism, is presently available on the Web, in many cases free of charge (e.g., CopyTracker®, Plagiarism. pro®).[4] This is not only useful for peer reviewers when evaluating manuscripts submitted for publication, but also in the academic world, to scrutinize undergraduate students' work, as well as MSc and PhD theses. That is the reason why printed theses submitted for public defense (viva voce) must always be accompanied by a PDF (or equivalent) version: in case of doubt, any element of the jury can upload part or the whole text and have it scrutinized by a software for plagiarism detection.

10.9.3   Promoting a Culture of Ethics

Besides other strategies to foster discipline, methodology, and dependability right from the very beginning of scientific education, *teaching ethics* to future, young, and not so young researchers is proving essential in the present context of the production of scientific knowledge. Quoting NAS et al. (1992), this is about "integrating ethics into the education of scientists" (see also NAS, 1995; Sigma Xi, 1986). In fact, there should be no excuse for any researcher to ignore his/her ethical rights and obligations within the field of scientific production, publication, and evaluation.

Teaching ethics may take different forms. It may be the main subject of a specific course, a conference, or a workshop or disseminated as an (important) aspect throughout a wide range of courses on different subjects. Regardless of the form adopted, ethics should not be taught as an abstract collection of general considerations and guidelines. Instead, a reflection based on case studies, addressing real situations, thus encouraging active participation and debate, is clearly a better strategy.

For example, if an ethics statement is viewed as a number of directives and "prescriptions" out of any practical context, and then a real situation occurs that is not directly covered in the set of written "prescriptions," the student will probably not know how to react. If, on the contrary, teaching is essentially focused upon the discussion of practical examples (including *ethical dilemmas*), using the ethics statement as a reference to recommend and foster good practices, then at the end of the course, the student will have acquired the necessary skills and autonomy to take the best decision in every real, practical context. In other words, teaching ethics is not about communicating a simple and abstract normative, regulatory framework: it should promote an ethics of responsibility, more than an ethics of statutory regulations.

---

[4] URL, in the same order: http://copytracker.soft112.com/download.html; http://plagiarism.pro/

10.9.4   *Reacting to Misconduct in Scientific Publication*

Despite the considerable effort of self-regulation that the scientific community makes nowadays in order to encourage its members to abide by ethical standards and practices, it would obviously be unrealistic to think that it is possible to completely abolish misconduct in science, and in particular in scientific publication. One question therefore arises: how to react to unethical behavior?

In practice, the feeling of impunity is actually an encouragement for any unethical behavior to be subsequently repeated, perhaps even in larger proportions, and leading to more serious consequences. That sentiment generally results from the absence of any sanction. The main role of a *sanction* is not to inflict a punishment, but to combat the feeling of impunity, therefore discouraging those deliberately responsible for questionable practices or misconduct from relapsing in the future. Any sanction should respect the *principle of proportionality*: (i) a more serious infraction leading to a more severe sanction; and (ii) equally severe sanctions corresponding to equally serious infractions. The work reported by the reflection panel of the National Academy of Sciences (NAS et al., 1992) is an excellent reference in this regard.

In fact, when facing any ethical deviation, one should follow three successive steps, in this order: (i) first, detect and interpret possible signs that may indicate that a problem is imminent, thus trying to avoid its occurrence; (ii) if that is not possible and a problem does occur, investigate it thoroughly, gathering all available elements that may prove useful, in order to figure out whether responsibility for misbehavior should be attributed and to whom; (iii) finally, and only then, apply sanctions to those found responsible.

In occurrences linked to scientific publication, different forms of sanctions may be applied. Typically, in increasing order of severity, a sanction may be: oral reprimand (possibly with the threat of a penalty); written reprimand (sometimes requiring a correction of the error and/or an official apology); systematic scrutiny of future activity; suspension or termination of financial support; obligation to return undeserved sponsoring; imposition of a fine; requirement to publish a clarifying note or to simply withdraw the faulty publication; exclusion from future panels of peer evaluation; downgrading of the contractual position; exoneration; temporary banning from attending the school (for students); dismissal (employees or collaborators); or expulsion (students). In any case, a sanction always means irreparable damage to the culprit's professional and personal image.

A final note is addressed to undergraduate students. It is very likely that you will be the researchers of the future. The number of cases of coursework that is plagiarized from the Internet has been increasing in the last decade. When detected, this type of fraud should be combated with zero tolerance, in order to avoid sending a wrong message that would simply encourage the sentiment of impunity. The transition from an "innocuous" plagiarized piece of coursework to a scientific publication is a smaller step than it might look at a first glance!

10.10   Conclusions

The aim of scientific publications is to report scientific research and disseminate the corresponding outcomes. They may be seen as having one of two main goals: making the research results and conclusions publically available for direct use in practical applications; or to be used by other researchers as a basis for further research. In both of these cases, there is a major characteristic that has to be fully respected: published results must be reproducible. If they are not, the corresponding publication is not only useless but also prejudicial: this is because, as it is essentially made merely of scientific noise, it is occupying the place of a more deserving one. In itself, just the condition of reproducibility implies compliance with other crucial values underpinning what any trustworthy publication should display, namely, quality, honesty, clarity, transparency, neutrality, verifiability, openness to debate and criticism, and responsibility. In other words, for a scientific publication to fulfill its role in building scientific knowledge, it, and the research it is based upon, must be ethically irreproachable. A publication is no exception to the general principle stating that quality and ethics are closely and directly linked: neither exists without the other.

Pressure to publish ("publish or perish") is perhaps the most influential factor that presently threatens the due respect for values in scientific publication. However, other forms of conflicts of interest and of commitment also originate questionable practices or even misconduct in science. This is apparent in several practical circumstances, namely, when it comes to assessing the quality of published scientific knowledge. Unethical behavior while conducting, reporting, and assessing scientific research is responsible for the occurrence of different types of errors, the consequences of which may range from simple inconveniences to life-threatening disasters. It is therefore of the upmost importance to consider and implement all possible strategies and mechanisms for avoiding or minimizing negative impacts. Those strategies include: ever intensifying and improving monitoring, regular scrutiny, and (if necessary) sanctioning; dissemination and permanent improvement of ethics statements; and, probably above all, promoting the teaching of ethics since the very beginning of the future scientists' education.

Conscience is crucial on the individual scale when producing and reporting science; confidence is decisive in the relationship uniting the scientific community and between scientists and society. In this regard, optimism is justified by reality: in the inevitable conflict between ethical values and the threats they are exposed to, ethics is noticeably gaining ground.

References

Amin M, Mabe M (2000) Impact factors: use and abuse. Perspectives in Publishing, no. 1: 1–6. Elsevier Science

Anderson M, Kleinert S (2013) Montreal statement on research integrity in cross-boundary research collaborations. Third world conference on research integrity. Montreal, Canada, 5–8 May 2013

Ashby M (2005) How to write a paper, 6.th edn. Cambridge University Press, Cambridge

Boileau-Despreaux N (1674) L'art poétique. Stéréotype d'Herhan, Paris

Elliot D, Stern JE (eds) (1997) Research ethics: a reader. University Press of New England, Hanover

Garfield E (2006) The history and meaning of the journal impact factor. JAMA 295(1):90–93

Harzing AW (2007) Publish or perish. http://www.harzing.com/pop.htm. Accessed at 2016

Hirsch JE (2005) An index to quantify an individual's scientific research output. Proceedings of the National Academy of Sciences (PNAS) 102(46):16569–16572

Lawrence PA (2003) The politics of publication: authors, reviewers and editors must act to protect the quality of research. Nature 422(6929):259–261

Marshall S, Green N (2007) Your PhD companion, 2nd edn. How To Books Ltd, Oxford

Mayer T Steneck N (2007) Final report. First world conference on research integrity: Fostering responsible research. Lisbon, Portugal, 16–19 Sept 2007

Mayer T, Steneck N (2010) Singapore statement on research integrity. Second world conference on research integrity. Singapore, 21–24 Jul 2010

NAS, NAE, IOM (1992) Responsible science: ensuring the integrity of the research process, vol I. Washington, DC, National Academy of Sciences/National Academy of Engineering Institute of Medicine/National Academy Press

NAS (1995) On being a scientist: responsible conduct in research, 2nd edn. National Academy of Sciences. National Academy Press, Washington DC

Oliveira LA (2012) Dissertação e tese em ciência e tecnologia, 2nd edn. Lidel, Portugal

Oliveira LA (2013) Ética em investigação científica. Lidel, Portugal

Parnas DL (2007) Stop the numbers game: counting papers slows the rate of scientific progress. Commun ACM 50(11):19–21

Stewart CN Jr (2011) Research ethics for scientists: a companion for students. Wiley-Blackwell, John Wiley & Sons, Ltd, Hoboken

Sigma Xi (1986) Honor in science, 2nd edn. Sigma Xi – The Scientific Research Society, New Haven

U.S. Government (1949) "the Nuremberg code", trials of war criminals before the Nuremberg military tribunals under control council law, vol 2. U.S. Government Printing Office, Washington DC10, pp 181–182

WMA 1964. Declaration of Helsinki – ethical principles for medical research involving human subjects. 18th World Medical association (WMA) General Assembly, Helsinki, Finland. PDF version available at the following Internet address: http://www.wma.net/en/30publications/10policies/b3/ (11 June 2015)

Zare RN (2012) Editorial: assessing academic researchers. Angew Chem Int Ed 51(30):7338–7339

Part IV
Food Ethics Case Studies

Chapter 11
Operation Aberdeen

Simon Dawson

11.1 Background

Denby Poultry Products (DPP) was opened on 28th August 1992 as a licensed knackers yard (HM Government 2015). The business, owned by Mr. Peter William Roberts, purchased old, rotten, diseased, or generally unacceptable quality chickens and turkeys for processing into waste and fertilizer. Under the Slaughterhouses Act 1974, the site had to be inspected annually by the local authority (LA) to ensure it was both sanitary and suitably managed for the disposal of the animals and their parts.

DPP remained like this for 5 years until 1997 when Mr. Roberts applied to the local council to become a pet food processor. The premises were inspected by local Trading Standards officers and granted registration by MAFF under the Animal By-Products Order 1992, which was updated to the Animal By-Products Order 1999, to process low-risk waste into pet food (Haslam 2011).

From 1997, Mr. Roberts worked with others collecting and processing unfit poultry meat waste, from licensed slaughterhouses into pet food. At that time, the Animal By-Products Order gave specific requirements and restrictions for the handling and processing of animal by-products to ensure that the products were not intended for human consumption (Animal By-Products Order 1992; Animal By-Products Order 1999). However, Mr. Roberts discovered the potential profit available if the unfit poultry meat were to be sold back to the human food supply chain.

This case study is an example to illustrate, in the absence of sensitivity to food ethics, how easy it can be to purchase, process, and sell unfit food.

S. Dawson (✉)
Cardiff Metropolitan University, Cardiff School of Health Sciences,
Llandaff Campus, Western Avenue, Cardiff CF5 2YB, UK
e-mail: sdawson@cardiffmet.ac.uk

© Springer International Publishing AG 2018
R. Costa, P. Pittia (eds.), *Food Ethics Education*, Integrating Food Science and
Engineering Knowledge Into the Food Chain 13,
https://doi.org/10.1007/978-3-319-64738-8_11

Licensed slaughterhouses paid on average £100 per ton for the disposal of poultry waste; therefore, when DPP offered to buy their waste for £25 per ton, it became a very tempting offer. MHS inspectors visit licensed slaughterhouses as a daily routine; however, their service-level agreements with the Department for Environment, Food and Rural Affairs (DEFRA) only allocated 15 min per month for the supervision of the storage, segregation, and disposal of animal by-products (Clark 2003). Due to this very limited time, the transactions with DPP were not highlighted as a cause for further investigation.

The poultry waste upon arrival at DPP was diseased, was rotten, and often had a putrid off-odor; therefore, selling this back into the food chain was not a straightforward task. Initially, DPP had to "dress" the poultry meat so that it appeared fresh. This was accomplished by soaking the unfit poultry in tanks of sodium hypochlorite solution. Sodium hypochlorite removed surface slime, typically caused by pseudomonads (Jay et al. 2005); deodorized the meat; removed some blemishes; and lightened and "bleached" its color, giving it a somewhat fresh appearance. The poultry was then dipped in a brine solution of salt and water, to remove the chlorine odor. High concentrations (above 200 ppm) of sodium hypochlorite on poultry meat can produce carcinogenic trihalomethanes such as chloroform in the meat (Arbor 2008; Vizzier-Thaxton et al. 2010). This was not considered during recycling or during the prosecution case that followed.

Once the poultry had been cleaned, it then required an EC health mark to show that the poultry had been produced in licensed premises to regulated hygienic standards. Gary Drewett, owner of a Northampton-based poultry food processor MK Poultry, worked with Mr. Roberts by both supplying DPP the unfit poultry meat and adding illegitimate European health marks to the recycled products (Beirne and South 2013).

At this time, the supplier approval process was fairly relaxed. In many cases simply showing a health mark with samples of suitably high-quality poultry meat was enough of an approval process to allow the products' rite of passage. Supplier approval audits involving site visits were uncommon. This meant that customer deception by DPP was simple as no customers visited the production facilities in Derbyshire prior to approving their products.

The three main routes of sale used by Denby Poultry Products to sell the unfit poultry include wholesale meat, licensed cutting plants, and market trade/wholesaler (MK Poultry). Fig. 11.1 is a photograph taken by Amber Valley Borough Council of the entrance to Denby Poultry Products production site. It highlights the dilapidated condition of the external construction of the premises and lack of doorways/segregation into the storage area.

11.1.1   Wholesale Meat

The largest customer base for DPP was identified as meat wholesalers in the North West England. A total of 311 potential customers had been highlighted, and from this, 128 contacts were established (Haslam 2011). As with market trade, caterers

Fig. 11.1 Entrance to Derby Poultry Products (Courtesy of Amber Valley Borough Council)

and butchers were identified as target customers along with local schools, hotels, market stalls, and restaurants, although none were publically displayed.

11.1.2 Licensed Low-Throughput Cutting Plant

A licensed low-throughput cutting plant in Northampton sold the unfit poultry to 105 identified customers. This achieved nationwide distribution due to the sales through food brokers, wholesalers, and cutting plant and food manufacturers. During the police raid, two-thirds of the collective total of unfit poultry were seized here. Following an investigation, Tesco, Sainsbury's, the former retailer Kwik Save, and Shippams Foods (owned by Princes Limited) were identified as DPP customers. All of which conducted product recalls totaling approximately £1 m after warnings through the Food Standards Agency about the potential contamination and food safety risks involved in DPP products (Anon 2003c; FSA 2001).

11.1.3 Market Trade and Wholesaler

Distribution through the market trade and wholesaler (MK Poultry) identified 14 markets/customers in total. Although there was limited wholesale distribution, there had been a nationwide market trade which included sales to both butchers and caterers (Haslam 2011). During the raid by police and health authorities, no products were found on the premises traced back to DPP.

11.2   Surveillance

Until the initial call to environmental health on 7th December 2000, authorities were unaware of the food fraud case currently taking place. Additional contacts were made with the informant, and following this, a multiagency meeting was held in Derbyshire on 14th December involving health officials from local authorities (LA), FSA, MAFF, MHS, TSO, and Derbyshire Police (Haslam 2011). Information was shared between parties, and boundaries were established as to who would take responsibility. At this point, there was an underlying need for confidentiality as well as accuracy in identifying authenticity and severity of issues raised.

Strict timescales were placed on gathering evidence as it had been seen as a public health priority. In February 2001 Derbyshire Police agreed to pursue this as an enquiry and launched a 32-day surveillance of the premises (Haslam 2011). During the observations, five deliveries to food outlets were recorded on video, to be later used as evidence. The operation progressed within the least amount of time (FSA 2001); however, sufficient evidence still had to be gathered to ensure a successful outcome.

11.3   The Raid

On 22nd March 2001, between 8 and 10 am, an organized operation led by Derbyshire Constabulary, Amber Valley Borough Council, and the FSA took place. This involved the raid of 20 premises, homes, and accountants by those thought to be involved in the alleged food fraud (FSA 2001; Haslam 2011).

A total of 150 police officers, 50 environmental health officers, five local authorities, and meat hygiene service personnel were involved in the operation. Denby Poultry Products had its pet food producer registration suspended by MAFF under the Animal By-Products Order 1999. Twenty-two police warrants were executed, and 16 people were initially arrested (FSA 2001), although six were later released.

Accompanying the raid at the Denby Poultry Products site was the BBC TV program "Panorama." The program described the putrid smell of rotten meat and showed footage of rats running around inside the premises and overflowing drainage leaking sewage over the floor. Judy Mallaber, former Minister of Parliament (MP) for Amber Valley presented at the House of Commons following the raid and stated "anyone who looks at the video footage or the pictures in my police file would just feel sick. It is disgusting" (Clark 2003).

Thirty tons of unfit poultry meat were seized during the raid. This included 10 tons from Denby Poultry Products and 20 tons from HK Poultry cutting plant. Chief Superintendent Fran Muldoon, of Derbyshire Constabulary, stated that "Today's actions are the culmination of four months of observations and operations. We have worked closely with Amber Valley Borough Council. The Food Standards Agency and other agencies have also been involved and this has been an excellent example of multi-agency working. The arrests we have made today do not mean the operation is complete. This is the initial stage of the operation. It is our intention that the

Fig. 11.2 Plastic dolavs containing unfit whole and part-processed chicken and turkey; 30 tons of meat were seized in total (Courtesy of Amber valley Borough Council)

investigation will try to trace the distribution chain of the poultry meat that is alleged to have been introduced into the human food chain and bring the operation to a conclusion" (FSA 2001).

The photographs shown overleaf (Figs. 11.2, 11.3, 11.4, and 11.5) were taken by Amber Valley Borough Council Environmental Health officers during the raid at the Denby production site.

Figures 11.2 and 11.3 highlight the unhygienic storage conditions of the raw poultry products and level of rancidity present in the products purchased by DPP. Figure 11.4 shows the prepared chicken breasts, after cutting highlighting the dirty, rusted racking, flooring, and damaged surrounding walls. The last photograph (Fig. 11.5) is the final prepared chicken breasts, after bleaching and packing, ready for sale. It is difficult to imagine that the unfit poultry taken from Figs. 11.2 and 11.3 would eventually become those shipped out in Fig. 11.5. Figure 11.6 provides a scheme of the supply chain.

11.4   Scale of Investigation

From the date of the first enquiry call in 2000 to the conviction of the offenders in 2003, the investigation lasted 3 years and cost the government approximately £1.75 m (Haslam 2011). Over £1 m was a result of police time alone (Clark 2003).

One hundred fifty police officers from five forces, along with 50 environmental health officers from five councils, were involved in the raid of the properties.

Fig. 11.3 Unfit rancid whole turkeys, ready to be prepared for sale (Courtesy of Amber Valley Borough Council)

Fig. 11.4 Prepared raw chicken breasts laid out on plastic trays ready for packing (Courtesy of Amber Valley Borough Council)

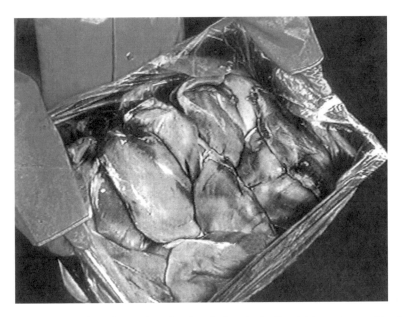

Fig. 11.5 Prepared chicken breasts boxed and packed ready for dispatch to customers (Courtesy of Amber Valley Borough Council)

These included Derbyshire (as lead and coordinator), Northamptonshire, West Yorkshire, Buckinghamshire, and Greater Manchester (FSA 2001).

Statements were taken from 329 individuals both as workers involved in the food fraud and customers receiving the goods. A total of 15,000 documents (1327 exhibits) were used, and 674 businesses were handled in the investigation (Haslam 2011).

Denby Poultry Products had been purchasing poultry waste for approximately £25 per ton and, after rework, had been selling it back into the human food supply chain for in excess of £1500 per ton (Muir 2003; Anon 2003a, b). This gives a markup of at least 5900%, netting DPP over £1 m profit, Mr. Roberts over £500,000, and his colleague Mr. Lawton, who took over the business in 2000, over £300,000 (Muir 2003).

11.5 Results of Trial

After a 12-week trial at Nottingham Crown Court, the crown prosecution service discharged three of the accused men as the jury failed to reach a verdict after a 36 h deliberation (Anon 2003a); the fourth was also discharged for his limited role within the case. Five others pleaded guilty to defraud, each of which received custodial sentencing totaling more than 15 years. The original owner of DPP Mr. Roberts was not present during the trial. Table 11.1 shows each of the accused, their position within the food fraud case, and sentencing passed by the crown prosecution service.

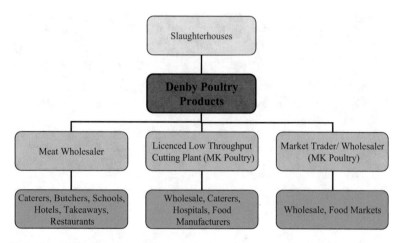

Fig. 11.6 The supply chain from primary processor to market used by Denby Poultry Products (Reconstructed from: Haslam 2011)

Table 11.1 Name of accused, position, and prison sentencing handed out

Name	Position	Sentencing
Peter William Roberts	Owner of DPP from 1992 to 1999	6 years
Robert Mattock	Owner of DPP from December 1999 to February 2000	2 years
David Lawton	Owner of DPP from February 2000 to March 2001	3 years and 4 months
George Allen	Worker at DPP	15 months
Gary Drewett	Owner of MK Poultry	2 years and 6 months
Mark Durrant	Manager at MK Poultry	12 months of suspended sentence

Reconstructed from Haslam (2011), FSA (2004) and Anon 2003c)

Police had been unable to locate Mr. Roberts following his 6-year prison sentencing as he fled the country prior to trial. As a result, a nationwide manhunt ensued trying to track him down. Four years passed before an anonymous call to news, reporters at BBC TV, who had been investigating his disappearance, stated that Mr. Roberts had settled in Northern Cyprus. BBC reporter Alistair Jackson, and an accompanying camera crew, eventually tracked him down and while he was out shopping announced their presence (Jackson 2014; East Midlands Today 2007; Anon 2008). The discovery of Mr. Roberts was presented to Derbyshire Constabulary. Detective Inspector Neil Perry from Derbyshire Police stated that "there is no extradition treaty with Northern Cyprus, it is not recognized by the British Government, therefore we have to use tact and diplomacy to inform the Turkish Authorities that they have a fugitive living on their territory" (East Midlands Today 2007). He was deported back to the UK in 2007 where he was taken to St. Mary's Wharf Police Station in Derbyshire. Mr. Roberts was then taken to Nottingham Crown Court where sentencing was passed. He was released in 2013.

11.6   Conclusions

At some point in the adulteration of the rotten poultry, those managing the process, those handling the products, and those adding false EC health marks had to know of their blatant disregard for human safety. Yet even with the unhygienic conditions of the production facility, rancidity of the poultry, and serious threat to cause harm, the recycling of the food waste was still carried out. Money clouded the minds of those involved in the food fraud case, and as a result, food ethics were disregarded.

11.7   How to Use This Case Study

One approach to using this case study would be to have the students read the case and then pose the following questions to them:

- What are the key facts within this case?
- Why do you think it took 3 years to investigate this case?
- How does this case encompass food ethics?
- What could been done to prevent an incident like this from occurring?
- What would you recommend going forward, and why?

An alternative approach, instead of posing questions as above, would be to set students into groups and have them role-play the part of the people involved in this case. Using this method will both actively engage students and allow them to understand the viewpoints of the case characters.

Acknowledgment We would like to thank Amber Valley Borough Council for their support in providing information and allowing their photographs to be used for this case study, along with Derbyshire Constabulary for the successful arrests and convictions of six of the ten defendants. In addition, a special thank is given to BBC reporter Alistair Jackson for tracking down Mr. Peter William Roberts in Northern Cyprus, who fled the UK before sentencing, allowing the British authorities to bring him to justice.

References

Animal By-Products Order (1992) SI 1992/3303. The Stationary Office, London
Animal By-Products Order (1999) SI 1999/646. The Stationary Office, London
Anon (2003a) Unfit meat jury discharged. BBC News. [Online]. Accessed 9 Jan 2015. Available from: http://news.bbc.co.uk/1/hi/england/derbyshire/3189099.stm
Anon (2003b). Illegal meat trade's 'drugs profits'. BBC News. [Online]. Accessed 9 Jan 2015. Available from: http://news.bbc.co.uk/1/hi/england/nottinghamshire/3190889.stm
Anon (2003c) Four jailed for poultry scam. BBC News. [Online]. Accessed 9 Jan 2015. Available from: http://news.bbc.co.uk/1/hi/england/derbyshire/3123414.stm

Anon (2008) British poultry supplier who was accused for selling rancid meal to schools and hospitals lives in the occupied areas. Republic of Cyprus Press and Information Office. [Online]. Accessed 3 Jan 2015. Available from:http://www.hri.org/news/cyprus/tcpr/2008/08-11-04.tcpr.html

Arbor A (2008) Benefits and risks of the use of chlorine-containing disinfectants in food production and food processing. Report of a joint FAO/WHO expert meeting. 27–30th May. World Health Organization, Geneva

Beirne P, South N (2013) Issues in green criminology – confronting harms against environments, humanity and other animals. Routledge, Oxon

Clark P (2003) Meat Fraud. House of Commons Hansard Debates. Part 31 [Online]. Accessed on 4 Jan 2015. Available from: http://www.publications.parliament.uk/pa/cm200304/cmhansrd/vo031210/debtext/31210-31.htm

East Midlands Today (2007) BBC One (September 15th)

Food Standards Agency (FSA) (2001) Multi-Agency operation into allegations of unfit meat entering the human food chain [Online]. Accessed on 3 Jan 2015. Available from: http://tna.europarchive.org/20110116113217/http://www.food.gov.uk/news/pressreleases/2001/mar/multiunfitmeat

Food Standards Agency (FSA) (2004) Denby illegal meat trial: lessons learned from "Operation Aberdeen" [Online]. Accessed on 3 Jan 2015. Available from: http://tna.europarchive.org/20130814101929/http://www.food.gov.uk/about-us/how-we-work/our-board/boardmeetoccasionalpapers/2004/paperinfo03_12_02

Haslam S (2011) Operation Aberdeen: and Beyond, What happens after the phone rings? Amber Valley Borough Council

HM Government (2015) Denby Poultry Products Limited 02743469. HM Government Companies House. [Online]. Accessed 8 Jan 2015. Available from: http://data.companieshouse.gov.uk/doc/company/02743469

Jackson A (2014) Maggot Pete – Behind the Scenes. BBC Inside Out. [Online]. Accessed on 2 Jan 2015. Available from: http://www.bbc.co.uk/insideout/content/articles/2007/08/20/east_midlands_11_maggot_pete_feature.shtml

Jay JM, Loessner MJ, Golden D (2005) Modern food microbiology, 7th edn. Springer, New York

Muir H (2003) Chicken racket highlights food flaws. The Guardian. [Online]. Accessed 1 Jan 2015. Available from: http://www.theguardian.com/uk/2003/aug/30/foodanddrink

Slaughterhouses Act (1974) c.3. The Stationary Office, London

Vizzier-Thaxton Y, Ewing ML, Bonner CM (2010) Generation and detection of trihalomethanes in chicken tissue from chlorinated chill water. The Journal of Applied Poultry Research 19(2):169–173

Chapter 12
Bushmeat

Simon Dawson

12.1 Background

The first published paper highlighting the term "bushmeat" originated within a journal in 1843 when the author Captain W. Allen accounted for his visit along the rivers of Cameroon in Western Africa. Allen (1843, p5) wrote, "the natives, who set fire to the grass in the dry season for the purpose of catching wild animals, which they call 'bushmeat'." Since then, the term bushmeat accounts for all "wild animals hunted for food, especially in Africa; the meat from these animals" (Oxford English Dictionary 2015).

In Central and Western Africa, bushmeat is essential for rural communities as a part of their diet, trade revenue, or culture. However, due to the continents unprecedented population growth, consumption and sale has risen to an unsustainable level that will result in the extinction of many species unless significant changes occur (Ape Alliance 1998; Chaber et al. 2010). The black rhinoceros (*Diceros bicornis*), western gorilla (*Gorilla gorilla*), and African wild ass (*Equus africanus*) are examples of mammals that are now critically endangered due to illegal poaching (Red List 2015a; Red List 2015b; Red List 2015c). A report published by UNICEF (2014) highlights that over the next 35 years, Africa will double in population with over 1.8 billion new baby births, indicating this problem is only going to get worse, particularly with improving life expectancy.

Over the last century, Western and Central Africans have migrated throughout the world, establishing communities in many major cities and towns (International Organization for Migration 2011). Within these countries, a variety of protein sources are readily available, yet consumers are willing to pay high prices in order

S. Dawson (✉)
Cardiff Metropolitan University, Cardiff School of Health Sciences, Llandaff Campus, Western Avenue, Cardiff, CF5 2YB, UK
e-mail: sdawson@cardiffmet.ac.uk

© Springer International Publishing AG 2018
R. Costa, P. Pittia (eds.), *Food Ethics Education*, Integrating Food Science and Engineering Knowledge Into the Food Chain 13,
https://doi.org/10.1007/978-3-319-64738-8_12

Fig. 12.1 Giraffe killed for meat; poachers arrested by Big Life Rangers (Courtesy of Big Life Foundation, photographer Nick Brandt)

to obtain bushmeat species native to their homeland. To make matters worse, the increased wealth of consumers means traders are willing to go to extreme lengths in order to export/import bushmeat due to the financial incentives available. Therefore, even the more resilient species may end up endangered, in order to supply this ever-growing demand (Chaber et al. 2010). The photograph (Fig. 12.1) below shows the remains of a giraffe slaughtered for bushmeat by two poachers, both of which were arrested (Big Life Foundation 2015). Tracker dogs are often used by bushrangers to hunt for traps, injured and dead animals, and poachers. Figure 12.2 is a photograph of the remains of wild animals loaded onto a truck, including a zebra caught in snares from poachers (Big Life Foundation 2015).

There are many concerns regarding bushmeat consumption including eradication of species (John et al. 2002), poor hygiene (Federal Department of Home Affairs, 2014), inhumane slaughtering (Humane Society International 2015), and spread of tropical diseases (FDHA 2014; Subramanian 2012; Greger 2007).

Bushmeat is predominantly obtained from areas of extreme poverty including Africa, parts of South America, and Asia. Data provided by the World Bank (2015) shows that in 2013, of the top 20 poorest countries in the world, 18 of these are in Africa. Within these countries, moral principles governing the activities of a person and their behavior are lost in the need for survival. As a comparison, 12 of the richest countries in the world are in Europe (World Bank 2015).

Criminal gangs have exploited the desperate need for survival by persuading natives in abject poverty to hunt wildlife for trade. Endangered species are seen as being premium catches, procuring much higher markup prices; therefore natives receive incentives for obtaining these, even though this is completely illegal (File on 4 2004).

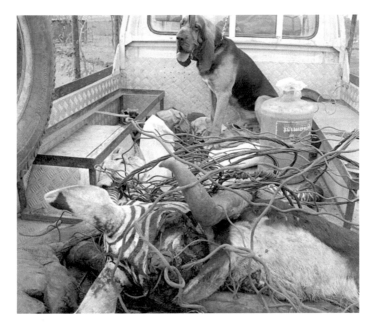

Fig. 12.2 Tracker dog with carcasses found in snares by Big Life Rangers (Courtesy of Big Life Foundation, photographer Nick Brandt)

12.2 Zoonotic Diseases

A zoonotic disease is one that is transferrable between animal and human host. These can be caused by any microorganism, including bacteria, parasites, viruses, fungi, and prions. Some of the most lethal diseases known to man have been the result of transference from animal to human, including anthrax, Ebola, and variant Creutzfeldt-Jakob disease (vCJD).

Non-domesticated animal species have an intricate role in the spread of endemic and emerging zoonotic diseases throughout the world. Since the turn of this century, international travel and migration, human encroachment and habitat expansion, and trade of live/dead animals have significantly increased (Travis et al. 2011). Endemic and emerging tropic diseases have also followed suit. Over the last 20 years, the media has headlined emerging diseases such as severe acute respiratory syndrome (SARS), Ebola virus disease (EVD), human immunodeficiency virus (HIV), and monkeypox, all of which originated from wild animals (Doyle 2015; Malone 2014; BBC News 2003).

Table 12.1 presents zoonotic diseases originated from wild animals, which have spread to humans through zoonosis. The microorganisms responsible for these diseases are not particularly heat resistant; therefore, thorough cooking is typically enough to destroy all of these. However, poor personal hygiene, cross-contamination

Table 12.1 Zoonotic tropical diseases spread from mammals to humans as bushmeat

Disease/Agent	Symptoms	Host animal(s)	Associated source(s) of origin
Ebola virus disease (including Ebola virus, Sudan virus, Reston virus, Bundibugyo virus, Tai Forest virus)	Fever, headache, muscle and joint pain, sore throat, internal bleeding, bleeding from eyes, nose, and mouth. Fatal in up to 90% of cases. No vaccines are currently available	Bats, primates, antelope, porcupine	Africa, predominantly Central and Western Africa
Leptospirosis	Fever, headache, nausea and vomiting, poor appetite, conjunctivitis, muscle pain, jaundice, symptoms of meningitis, seizures, coughing up blood. Can lead to brain damage, kidney failure, and death	Turtles/ Tortoises/ Reptiles	South America
		Mongoose	Africa, predominantly Central and Western Africa
Monkeypox virus	Fever, rash, intense headache, lymphadenopathy, myalgia. Can be fatal in young children	Primates, rodents	Africa, predominantly Central and Western Africa
Orthohepevirus A/ hepatitis E virus	Jaundice, anorexia, abdominal pain, nausea, vomiting, fever, hepatomegaly. Can result in acute liver failure which can lead to death	Deer, wild boar	Canada, UK, Europe, Asia, North America, South America
Rabies	Fever, paresthesia, paralysis, inflammation of the brain and spinal cord, coma. Can be fatal	Kudu, marmosets	Africa, predominantly Central and Western Africa
		Raccoon, squirrel	North America
Severe acute respiratory syndrome (SARS)	Fever, fatigue, headaches, chills, diarrhea, breathing difficulties. Reduction of oxygen in the blood which can be fatal	Bats, civets	Southern China
Simian foamy virus (SFV)	No reported illnesses from people tested positive for SFV	Primates	Africa, predominantly Central and Western Africa
Simian immunodeficiency virus/human immunodeficiency virus (HIV)	Nausea, vomiting, fever, enlarged lymph nodes, muscle aches, skin rash, weight loss. No vaccines are currently available; however, over the last 20 years, antiretroviral treatments have significantly improved life expectancy	Primates	Africa, predominantly Central and Western Africa

(continued)

Table 12.1 (continued)

Disease/Agent	Symptoms	Host animal(s)	Associated source(s) of origin
T-cell lymphotropic virus-1/human T-cell lymphotropic virus (HTLV)	Progressive weakness, muscle spasms, stiffness in muscles, constipation, weak bladder control. 95% of infected are asymptomatic. No vaccines are currently available	Primates,	Africa, Australia, North and South America, parts of Northern Europe
Yellow fever virus	Fever, nausea, loss of appetite, muscle pains, headaches, liver damage, jaundice, bleeding in urine, kidney damage	Primates	Africa, predominantly Central and Western Africa

Reconstructed from Travis et al. (2011), National Science Foundation (2013), FDHA (2014), Falk et al. (2013), Smith et al. (2012), WHO (2015) and WHO (2014)

during preparation, cuts, and exposed wounds are known routes that have allowed human infection. The handling and preparation of infected wildlife is therefore one of the critical transmission routes that must be carefully controlled. Table 12.2 highlights examples of highly resistant diseases originated from wild animals, including those associated with bushmeat. The *Bacillus anthracis* spores, responsible for anthrax disease, are resistant to sterilization temperatures, drying, and many disinfectants. These spores can easily be spread by release in the air, which is why the US Military has viewed it as potential biological terrorism threat (Inglesby 1999; FDA 2015).

12.3   Chemical Poisoning

Poisoning is widely used in many parts of Africa as a means of killing wild animals. The chemicals are cheap, easily accessible, silent, and very effective (Ogada 2014). Although illegal, they are often used to kill large carnivores such as lions, hyenas, and jackals as revenge attacks for killing farm animals. Poisons are also used to kill wild animals for bushmeat.

In 2001–2002, samples of bushmeat were taken and analyzed by scientific officers of the Ghana Standards Board. They found that 30% of samples contained chemical poisoning; all of which originated from locally sourced ingredients including residues of organochlorine, organophosphorus, and carbonates common in agricultural pesticides (FAO 2004). Further to this, Conservation International Ghana found chemical poisoning to be the second most popular method of hunting in Ghana (Opare-Ankrah 2007).

During a case study in 2007, involving traders and hunters from the Mfantseman District in Ghana, traders were asked about the severity to human health from the use of chemical poisons (Opare-Ankrah 2007). A general consensus was that no trader would buy or sell bushmeat that was believed to have been killed by poison.

Table 12.2 Highly resistant zoonotic tropical diseases spread from mammals to humans as bushmeat

Disease/Agent	Symptoms	Host animal(s)	Associated source(s) of origin
Bacillus anthracis	Symptoms differ depending on route of infection, however typically involve flu-like symptoms, nausea, vomiting, abdominal pain, headache, fever, swollen neck, bloody diarrhea, difficulty swallowing, meningitis, organ failure	Horse, cattle, sheep, camels, oxen, goat, donkeys, pigs, elephants, hippopotamus, lions, jackals, hyenas	Africa, North America, South America, Spain, Italy, Asia, Middle East, Philippines
Mycobacterium bovis/tuberculosis (TB)	Prolonged coughing, cough up blood, chest/breathing pains, chills, fever, fatigue, weight loss, seizures. TB is fatal if left untreated	Cattle, badgers, ferrets, cats, deer, lamoids	South America, North America, Africa, UK, Europe, Asia

. Reconstructed from Inglesby et al. (1999), FDA (2015) and Zumla et al. (2013)

One trader stated that, "people will not buy from me anymore if… get sick when they eat from here so I have to make sure the meat is good" (Opare-Ankrah 2007). This is easier said than done though as animals brought in for sale by hunters are often shot after being poisoned to prevent suspicion.

12.4   Bushmeat Trade

Due to the covert nature of the business and the extent of illegal international trade, it is impossible to give a true estimation of global bushmeat distribution (Smith et al. 2012). What is known though is that almost three-quarters of emerging diseases are of zoonotic origin, predominantly due to contact with non-domesticated animal species (Federal Department of Home Affairs 2014; Falk et al. 2013). The Parliamentary Office of Science and Technology (2005) estimated that between 1 and 3.4 million tons are harvested annually from Africa alone. This includes approximately 28 million bay duiker antelopes (*Cephalophus dorsalis*) and over 7 million red colobus monkeys (*Procolobus badius*).

Meat, milk, and their products are banned from entering the European Union (EU) from non-EU countries, even if imported in small quantities for personal consumption (EC regulation 745/2004). There are certain exceptions to this; however, with countries endemic with diseases such as foot and mouth disease (FMD) including Africa, the Middle East, and parts of South America (FAO 2012), the law is strict. In addition to this, many wild animals are regulated under the Convention on International Trade in Endangered Species of Wild Fauna and Flora (CITES) (CITES 2015a), an agreement which has been accepted throughout Europe (CITES 2015b).

In spite of the abovementioned regulations and widespread media coverage relating to bushmeat, thousands of tons are illegally imported in to Europe every year (Chaber et al. 2010). In France, for example, it is estimated that approximately 5 tons of bushmeat are smuggled from Africa through Paris Roissy Charles de Gaulle Airport every week in personal luggage (FDHA 2014). Of this, a third is estimated as being protected under CITES (FDHA 2014). It can be almost impossible to identify all meat products without DNA testing due to the similarity in appearance of meat cuts, especially if bones, skin, or hair is removed (Food Inspectors 2014); therefore, this figure may be much higher.

12.5   Bushmeat Cases in the UK

In November 1999, two shopkeepers were caught trading bushmeat from their store in Ridley Road Market in Dalston, East London (Vasagar 2001a). An environmental health officer, on a routine store visit, noticed a sign advertising bushmeat for sale (File on 4 2004). The price list contained CITES listed endangered species including tiger (*Panthera tigris*) and gorilla (*Gorilla gorilla*), both of which are illegal to hunt worldwide (CITES 2014). Mobalaji Osakuade and his partner Rosemary Kinnare told an undercover environmental health officer that they could get anything they wanted providing they were willing to pay for it. This includes whole lions at £5000, lion and tiger heads at £1000, and antelopes, goats, cane rats, porcupines, and live giant snails. The couple had also been smuggling illegal snakes and lizards, traditionally used within traditional African medicines. Each was given jail sentences of 4 months (Vasagar 2001a; Vasagar 2001b).

In December 2001, Dr. Yunes Teinaz, an environmental health officer for London Borough of Hackney, began an investigation into a shop in Kejetia Mini Market on West Green Road, London (Pointing and Teinaz 2004). During a routine inspection, Dr. Teinaz discovered meat products being prepared and sold in squalid, unhygienic, pest-infested conditions. The kitchen preparation area was so bad that an emergency prohibition notice was issued due to an imminent risk of injury to human health (Pointing and Teinaz 2004). Figures 12.3, 12.4, and 12.5 were taken by Dr. Teinaz during the inspection for evidence in court. The store was closed down and unable to reopen until substantial improvements warranted re-inspection and approval. However, Paulina Owusu Pepra, the store owner, and her partner had reopened without approval. On 22nd October 2002, Dr. Teinaz returned with police and found meat products again being prepared and sold in similar unhygienic conditions (Pointing and Teinaz 2004). Police officers seized over 2 tons of rotten bushmeat including cane rat, giant snails that were decomposing in their own feces, pigs' feet, and a range of unidentified smoked meat, some with skin still attached. The owner, Paulina Owusu Pepra, appeared in court in December 2003 and was given a 3-month prison sentence and lifetime ban for preparing food for sale. Her husband, believed to have fled the country before sentencing, has not been found (Pointing and Teinaz 2004).

Fig. 12.3 Oil and grime encrusted pot used for frying meat (Courtesy of Dr. Yunes Teinaz)

Fig. 12.4 Unknown smoked bushmeat prepared in dirty sink (Courtesy of Dr. Yunes Teinaz)

In 2004, the British Broadcasting Corporation (BBC) launched a program on BBC Radio 4 about the potential illegal bushmeat trade in London markets (File on 4, 2004). In the radio report, a senior environmental health officer stated "we have found it [bushmeat] on sale to some extent or another in almost every West African

Fig. 12.5 Deep-fried bushmeat left to cool on a paper plate (Courtesy of Dr. Yunes Teinaz)

shop in the area [Hackney]. We were finding forty, fifty kilos at a premises at a time. You could go back a month later and see exactly the same amount again. It's huge business." (File on 4 2004)

In 2012, the BBC received reports that shops in Ridley Road Market, London, were being used as distribution points for illegal meat (BBC News 2012; Lynn 2012). Undercover reporters, with hidden video cameras, were sent in to investigate further. During the report, Dr. Yunes Teinaz stated "this is providing meat in to the human food chain which can carry infectious diseases. The people who are arranging this illicit trade are very dangerous. They only observe financial gains" (BBC News 2012). The undercover reporters enter a few shops, and after discussions with some butchers, one states "you tell me one day before, then you can have it OK? I can't keep it here too much. Don't tell anyone… Otherwise there will be trouble you know" (BBC News 2012).

Two years later, a UK television documentary, the Food Inspectors, conducted a follow-up investigation in to London markets looking for illegal meat (Food Inspectors 2014). In one video, a butcher, who sells a reporter bushmeat, tells the under reporter how it enters the country. He states "it's coming under the table. Special. Africa, France, coming Dover under… underground" (Food Inspectors 2014).

12.6   Conclusions

Within the UK, the Products of Animal Origin (Third Country Imports) (England) Regulations 2006, as amended by the Products of Animal Origin (Third Country Imports) (England) (Amendment) Regulations 2007, prohibits the importation and sale of any meats outside of the European Union. However, with the strict regulations, government inspections, heavy fines, and potential jail sentences, this has not deterred the demand for this product.

It is expected that in extreme poverty-stricken areas, such as parts of Africa, South America, and Asia, bushmeat consumption is a necessity.

If wildlife is slaughtered only for these groups of people, and endangered species are left alone, this would become a sustainable option for the future. However, the majority of the World including the UK and Europe do not fall within this poverty-stricken category. Is bushmeat therefore essential to people living within these countries? No. Are ethical considerations made when bushmeat is illegally imported, sold, and consumed? Probably not. Therefore, before the endangered become extinct, we need to decide on how to act for the future. Instead of buying products that remind us of where we have come from, should we not be thinking more ethically and buy products that allow us to sustain for the future before it is too late?

12.7   How to Use This Case Study

One approach to using this case study would be to have the students read the case and then pose the following questions to them:

- How does this case encompass food ethics?
- What do you think drives bushmeat consumption?
- What are the risks involving bushmeat? [Think ethics as well as safety]
- What could been done to control/reduce bushmeat consumption?
- What would you recommend going forward, and why?

An alternative approach, instead of posing questions as above, would be to set students in to groups and have them role-play the part of the people involved in this case. Using this method will both actively engage students and allow them to understand the viewpoints of the case characters.

Acknowledgements We would like to thank Nick Brandt and Big Life Foundation for allowing us to use their photographs (Figs. 12.1 and 12.2). We would also like to thank Dr. Yunes Teinaz and, London Barrister, John Pointing for their assistance in Sect. 12.4 bushmeat cases in the UK and for permission to use their photographs (Figs. 12.3, 12.4 and 12.5).

Legislation   Commission Regulation (EC) No 745/2004 of 16th April 2004 laying down measures with regard to imports of products of animal origin for personal consumption.

The Products of Animal Origin (Third Country Imports) (England) Regulations 2006, as amended by The Products of Animal Origin (Third Country Imports) (England) (Amendment) Regulations 2007.

References

Allen W (1843) Excursion up the River of Cameroons and to the Bay of Amboises. London. Roy Geol Soc 13(1843):15

Alliance A (1998) The African Bushmeat trade – a recipe for Extinction. Fauna and Flora International, Cambridge

BBC News (2003) Europe plans SARS response. [Online] (last accessed 24 June 2015). Available from:http://news.bbc.co.uk/1/hi/world/europe/3003345.stm

BBC News (2012) BBC One (September 17th)

Big Life Foundation (2015) Photos from Big Life Kenya. [Online] (last accessed 27 June 2015). Available from: https://biglife.org/photos-videos/photos-from-big-life-kenya

Chaber AL, Allebone-Webb S, Lignereux Y, Cunningham AA, Rowcliffe JM (2010) The scale of illegal meat importation from Africa to Europe via Paris. Conserv Lett 3(2010):317–323

CITES (2015a) What is CITES? [Online] (last accessed 24 June 2015a). Available from:http://www.cites.org/eng/disc/what.php

CITES (2015b) List of Contracting Parties [Online] (last accessed 24 June 2015). Available from: http://www.cites.org/eng/disc/parties/chronolo.php

CITES (2014) Index of CITES Species. UNEP World Conservation monitoring centre, Cambridge

Doyle M (2015) The hunters breaking an Ebola ban on bushmeat. [Online] (last accessed 19 June 2015). Available from: http://www.bbc.co.uk/news/world-africa-31985826

Falk H, Dürr S, Hauser R, Wood K, Tenger B, Lörtscher M, Schüpbach-Regula G (2013) Illegal import of bushmeat and other meat products into Switzerland on commercial passenger flights. Rev Sci et Tech (Int Off Epizoot) 32(3):727–739

Federal Department of Home Affairs (FDHA) (2014) Bushmeat: information and identification guide. A collaboration of the Federal Food Safety and Veterinary Office and Tengwood Organisation, Federal Department of Home Affairs FDHA

File on 4 (2004) British Broadcasting Corporation Radio 4 Transcript 26th October 2004. [Online] (last accessed 20 June 2015). Available from: http://news.bbc.co.uk/nol/shared/bsp/hi/pdfs/fileon4_20041019_bushmeat.pdf

Food and Agricultural Organisation (FAO) (2004) The Bushmeat Crisis in Africa: conciliating food security and biodiversity conservation in the Continent [Online] (last accessed 26 June 2015). Available from: ftp://ftp.fao.org/unfao/bodies/arc/23arc/J1457e.doc

Food and Agricultural Organisation (FAO) (2012) Foot-and-mouth disease: Frequently asked questions. FAO/OIE Global Conference on Foot and Mouth Disease Control. Bangkok, Thailand 27–29th June

Food and Drug Administration (FDA) (2015) Anthrax. [Online] (last accessed 25 June 2015). Available from: http://www.fda.gov/BiologicsBloodVaccines/Vaccines/ucm061751.htm

Food Inspectors (2014). BBC One. (August 12th)

Greger M (2007) The human/animal interface: Emergence and resurgence of zoonotic infectious diseases. Crit Rev Microbiol 33(4):243–299

Humane Society International (2015) Get the facts about the wildlife trade. [Online] (last accessed 23 June 2015). Available from: http://www.hsi.org/issues/wildlife_trade/wildlife_trade_info-graphic.html

Inglesby TV, Henderson DA, Bartlett JG, Ascher MS, Eitzen E, Friedlander AM, Hauer J, McDade J, Osterholm MT, O'Toole T, Parker G, Perl TM, Russell PK, Tonat K (1999) Anthrax as a biological weapon: medical and public health management. Am Med Assoc 281(18):1735–1745

International Organization for Migration (IOM) (2011) Guide to enhancing migration data in West and Central Africa. International Organization for Migration, Geneva

John EFA, Currie D, Meeuwig J (2002) Bushmeat and food security in the Congo Basin: Linkages between wildlife and people's future. Environ Conserv 30(1):71–78

Lynn G (2012) Cane Rat Meat 'Sold to public' in Ridley Road market. BBC News, London

Malone A (2014) Secret trade in monkey meat that could unleash Ebola in UK: How an appetite for African delicacies at British market stalls may spread killer virus. Mail online. [Online] (last accessed 19 May 2015). Available from: http://www.dailymail.co.uk/news/article-2713707/Secret-trade-monkey-meat-unleash-Ebola-UK-How-appetite-African-delicacies-British-markets-stalls-spread-killer-virus.html

National Science Foundation (2013) Human disease leptospirosis identified in new species, the Banded Mongoose, in Africa [Online] (last accessed 24 June 2015). Available from: http://www.nsf.gov/mobile/news/news_summ.jsp?cntn_id=127914

Ogada DL (2014) The power of poison: pesticide poisoning of Africa's wildlife. Ann NY Acad Sci 1322(2014):1–20

Opare-Ankrah Y (2007) The Bushmeat trade, livelihood securities and alternative wildlife resources. In: A Case Study of Mankessim and its environs in the Mfantseman District (Ghana). Norwegian University of Science and Technology, Trondheim

Oxford English Dictionary (2015) Bushmeat. [Online] (last accessed 23 June 2015). Available from: http://www.oed.com/view/Entry/25179?redirectedFrom=bush+meat#eid11637187

Parliamentary Office of Science and Technology (2005) Postnote: the Bushmeat Trade. (No. 236) [Online] (last accessed 25 June 2015). Available from: www.parliament.uk/briefing-papers/post-pn-236.pdf

Pointing J, Teinaz Y (2004) Halal Meat and Food Crime in the UK [Presentation]. International Halal food seminar, Islamic University College of Malaysia

Red List (2015a) Diceros bicornis. [Online] (last accessed 23 June 2015). Available from: http://www.iucnredlist.org/details/6557/0

Red List (2015b) Gorilla gorilla. [Online] (last accessed 23 June 2015). Available from: http://www.iucnredlist.org/details/9404/0

Red List (2015c) Equus africanus. [Online] (last accessed 23 June 2015). Available from: http://www.iucnredlist.org/details/7949/0

Travis, DA., Watson RP., Tauer A. (2011) The spread of pathogens through trade in wildlife. Scientific and Technical Review of the Office International des Epizooties. 30(11): 219–239.

Smith KM, Anthony SJ, Switzer WM, Epstein JH, Seimon T, Jai H, Sanchez MD, Huynh TT, Galland GG, Shapiro SE, Sleeman JM, McAloose D, Stuchin M, Amato G, Kolokotronis S-O, Lipkin WI, Karesh WB, Daszak P, Marano N (2012) Zoonotic Viruses Associated with Illegally Imported Wildlife Products. PLoS One 7(1):1–9

Subramanian M (2012) Zoonotic disease risk and the bushmeat trade: Assessing awareness among hunters and traders in Sierra Leone. EcoHealth 9(4):471–482. PLoS ONE. 7 (1). pp 1-9

UNICEF (2014) Generation 2030: Africa. Division of Data, Research and Policy. UNICEF, New York

Vasagar J (2001a) Couple sold rare species as bushmeat court told. [Online] (last accessed 21 June 2015). Available from: http://www.theguardian.com/uk/2001/may/15/london

Vasagar J (2001b) Monkey meat dealers guilty of smuggling. [Online] (last accessed 21 June 2015). Available from: http://www.theguardian.com/uk/2001/may/26/jeevanvasagar

WHO (2014) Yellow Fever. [Online] (last accessed 26 June 2015). Available from: http://www.who.int/mediacentre/factsheets/fs100/en/

WHO (2015) Ebola virus disease. [Online] (last accessed 24 June 2015). Available from: http://www.who.int/mediacentre/factsheets/fs103/en/

World Bank (2015) GDP Per capita (current US$). [Online] (last accessed 26 June 2015). Available from: http://data.worldbank.org/indicator/NY.GDP.PCAP.CD?order=wbapi_data_value_2013+wbapi_data_value+wbapi_data_value-last&sort=asc

Zumla A, Raviglione M, Hafner R, von Reyn F (2013) Current Concepts: Tuberculosis. Review article. N Engl J Med 368(8):745–755

Chapter 13
Food Labelling Case Studies

Vitti Allender

13.1 Introduction

The term *ethical* is defined as relating to moral principles (Oxford English Dictionary 2014). Labelling of foods is at the very heart of consumer choice, and the Kantian concepts of rights and duties rather than costs and benefits (Mephram 2000) suggest that full and clear labelling information should be given to consumers irrespective of business goals.

The purpose of this chapter is to consider two case studies in which the moral acceptability of the labelling of food, from the perspective of the consumer, is examined. The chapter does not consider environmental and social labels in the food sector (e.g. Organic, Fairtrade,[1] Rainforest Alliance[2]). It will not cover controversial technologies that have given rise to rejection by the consumer such as the *Flavr Savr* tomato,[3] since it has already been established that moral and ethical concerns can affect level of acceptance of such technologies (Rollin et al. 2011). Finally, labelling concerns connected with food scares such as the Europe-wide *horse meat incident*[4] are not covered as these are well documented elsewhere (Elliott 2014). Instead, the cases will concentrate on everyday food products where the label serves not only to inform the consumer of the true nature of the food but also acts as a marketing tool.

---

[1] Fairtrade Labelling Organizations International.

[2] Rainforest Alliance/Sustainable Agriculture Network.

[3] The first genetically modified food – 1994.

[4] Foods purporting to contain other meats actually contained undeclared or improperly declared horse meat.

V. Allender (✉)
Cardiff Metropolitan University, Cardiff School of Sport and Health Sciences, Llandaff Campus, Western Avenue, Cardiff CF5 2YB, UK
e-mail: vallender@cardiffmet.ac.uk

© Springer International Publishing AG 2018
R. Costa, P. Pittia (eds.), *Food Ethics Education*, Integrating Food Science and Engineering Knowledge Into the Food Chain 13,
https://doi.org/10.1007/978-3-319-64738-8_13

221

Consumers are dependent on labelling information to make informed decisions, and the application of moral principles is part of this decision. The provision of healthier options by food manufacturers is driven by customer demand, public sector concerns and corporate body wish to be socially responsible (Lähteenmäki 2013). Although health and nutrition claims are closely regulated in the EU and giving false or misleading information on food labels is prohibited, truthful labels relying on supermarket ranges offering more healthy versions of products sometimes sit in a subjective area. Examination of a fruit juice drink and a 'healthier' version in case study 1 will explore these subjective areas.

With the exception of wines and spirits which must be packed in specified quantities in the EU, manufacturers are free to select the size and description of packages and serving sizes (Chandon and Wansink 2012). Hence, descriptors such as 'value', 'jumbo' and 'premium' can be regarded as promotional statements only that express the subjective views of the company. However, when faced with a range of similar products under the same branding, where each has slight differences in size, ingredients and price, there is cause for consumer confusion. This is compounded where consumers are faced with unfamiliar ingredients and processing techniques to the point that labels may simply be incomprehensible and hence can be considered to be unethical (van der Merwe and Venter 2010). Case study 2 will examine a range of hot dog products in order to consider the implications of having a large range of similar products and what may be understood by the consumer.

13.2   Case Study 1: JuicyCo Cranberry Juice Drinks

JuicyCo is a European drinks manufacturing company based in the UK. It produces a range of fruit juices, fruit nectars and fruit juice drinks in 1 liter rectangular cuboid composite cartons. The composition of fruit juices and fruit nectars is controlled in Europe under Union provisions,[5] with fruit juices being defined as 'the fermentable but unfermented product obtained from the edible part of fruit … having the characteristic colour, flavour and taste typical of the juice of the fruit from which it comes[6]'. *Cranberry juice* is not produced since pure juice from this fruit would be unpalatable. The company produces *cranberry nectar* for mainland Europe. Fruit nectar is the fermentable but unfermented product that is obtained by adding water to a juice along with sugars and/or honey. In order to be called *cranberry nectar*, there is a required[7] minimum juice content of 30%, and in the case of the JuicyCo product, this minimum is met together with the addition of water and sugar. For the UK market (where the presence of fruit nectars in retail establishments is relatively uncommon), the company produces two varieties of cranberry juice drink. The first

---

[5] Council Directive 2001/112/EC of 20 December 2001 relating to fruit juices and certain similar products intended for human consumption last amended by Directive 2012/12/EU.

[6] Annex 1 of Council Directive 2001/112/EC.

[7] This is the legal minimum as per Annex IV of Council Directive 2001/112/EC.

Cranberry juice drink with cranberry juice from concentrate

Ingredients: Water, Cranberry Juice from concentrate (10%), Glucose-Fructose Syrup, Malic Acid, Colour (Anthocyanins), Flavourings, Vitamin C

Nutrition	Per 100ml	Per 250ml
Energy	220kj 55kcal	545kj 130kcal
Fat	>0.1g	>0.1g
of which saturates	>0.1g	>0.1g
Carbohydrate	12.3g	30.8g
of which sugars	12.3g	30.8g
Protein	>0.1g	>0.1g
Salt	>0.01g	>0.01g
Vitamin C	32mg 40% NRV	80mg 100% NRV

Pack contains 4 servings

Refrigerate once opened and use within 5 days

JuicyCo Ltd
Cardiff
CF1 1AA

JuicyCo

Cranberry

Juice drink

1litre ℮

Best before end June 2017

Each 250 ml serving contains

Energy	Fat	Saturates	Sugars	Salt
545kJ 130kcal	>0.1g	>0.1g	30.8g	>0.01g
6%	> 1%	> 1%	34%	> 1%

% of your reference intake
Typical values per100ml: Energy 220kJ/55kcal

Fig. 13.1 JuicyCo's *Cranberry juice drink with cranberry juice from concentrate*

is *Cranberry juice drink with cranberry juice from concentrate*, and the second, which is marketed as part of the company's *Better Health* range of products, is *No added sugar cranberry juice drink with sweeteners*. Since *Cranberry juice drink* is not a reserved description,[8] there is no minimum cranberry content for these products.

Figure 13.1 shows the label for JuicyCo's *Cranberry juice drink with cranberry juice from concentrate*. The label shown represents the front face of the carton and the left-hand edge, folding at the dotted line.

---

[8]A reserved description relates to the name of a product which may only be used if that product conforms to particular compositional requirements. Fruit juices and fruit nectars are examples of reserved descriptions.

The label contains the mandatory particulars required by the EU Regulation 1169/2011 on Food Information to Consumers (FIC). The name of the food, *Cranberry juice drink with cranberry juice from concentrate*, is descriptive and is sufficiently clear to enable the consumer to know the true nature of the product and distinguish it from other products with which it might be confused.[9] However, there may be consumer confusion as to a product called a *juice drink* since it sits alongside other juice drinks and juices in the supermarket, often in chilled cabinets, and without studying the label, it is not easy to distinguish between the different products. The list of ingredients in descending order of weight[10] indicates 10% cranberry juice from concentrate[11] and lists the additive category 'colour' and the name of the colour.[12] It also lists 'flavourings' in which case specific flavourings do not need to be labelled.[13] With only 10% cranberry juice, it is evident that some of the flavour of cranberry comes from artificial flavours. The product contains malic acid. This is used to impart tartness to the product (Hui 2005). This tartness is characteristic of cranberries, but the low cranberry content is not sufficient to impart this quality to the drink, and so it must be characterized through malic acid. The net quantity of the food is given (1l) with the 'e' symbol indicating that the product has been packed in accordance with the European Average Weight System. The durability indication is given in the form of the best before date, which is appropriate since after a short period from this date, from a microbiological point of view, the product is not likely to constitute an imminent danger to human health.[14] Storage conditions are given along with the name and address of the food business operator.[15] Finally, a tabulated nutrition declaration is provided in the correct format.[16] Vitamin C content is highlighted in the nutrition table, and it is clear from the ingredients list that vitamin C has been added to the product as a separate ingredient. Consumers are given sufficient information to realize that the vitamin C content is not present as a result of the cranberries contained in the product but from another source.

The pictorial representation of cranberries on the front of the pack is appropriate, and shows luscious, juicy-looking cranberries. Underneath this picture is the optional UK front-of-pack (FoP) labelling. FoP labelling is permitted since Article 30 of FIC allows repeating of the energy value together with the amounts of fat, saturates, sugars and salt on the labels of prepacked foods. The FoP nutrition label aims to give consumers a quick reference point in respect of the content of these nutrients per 100 g/ml and/or per portion of the product. Threshold amounts trigger optional colour coding as additional forms of expression, with green, amber and red

---

[9] Articles 17(1) and 2 (1)(p) of FIC.

[10] Article 18 of FIC.

[11] Article 22 of FIC.

[12] Article 18 and Annex VII Part C of FIC. Either the name or the e number may be declared.

[13] Annex VII Part D of FIC.

[14] Article 24 of FIC.

[15] Article 9 of FIC.

[16] Article 30 of FIC – this is a voluntary declaration for this product which becomes mandatory on 14 December 2016.

representing a prominent visual descriptor for low, medium and high values of fat, saturates, sugars and salt. In the case of JuicyCo's *Cranberry juice drink*, the company has chosen not to use the colour coding. Whereas the low amounts of fat, sugars and salt would trigger green descriptors, the sugar content would mean a red colour code, detracting from what consumers would otherwise view as a healthy drink (Temple 2014).

Figure 13.2 shows JuicyCo's *No added sugar cranberry juice drink with sweeteners*. The provision of mandatory particulars is in line with the previous product, with some significant differences. FIC requires food products containing sweeteners to bear this declaration within the name of the food.[17] The ingredients list declares the added sweeteners, with the presence of aspartame triggering the statement 'contains a source of phenylalanine'.[18] The other difference in the ingredients list is that vitamin C has been replaced by the term *ascorbic acid.*

The front of the pack looks very different to the previous product. The light blue background, use of *Better Health Range* branding and the 'exercising woman' logo give a different image. The picture of cranberries used has green leaves, giving the consumer the impression that this is a healthier product. Indeed, the optional front-of-pack nutrition label flags green throughout, indicating low amounts of fat, saturates, sugars and salt. The three stars on the pack offer additional information for quick reference. Ten calories per serving is borne out by the nutrition declaration, and the fact that the product has no added sugar is emphasized. Finally, the product makes a health claim[19] that it is high in vitamin C. This claim is permitted since *high in vitamin C* requires the product to contain 15% of the Nutrient Reference Value (in this case 80 mg) in 100 ml of the product. The 24 mg supplied by 100 ml of *No added sugar cranberry juice drink with sweeteners* meets this requirement.[20]

The vitamin C claim on the product is interesting. The content is 60 mg per portion, offering 75% of the Nutrient Reference Value (NRV). In the ingredients list, there is no *vitamin C* declared, since the term *ascorbic acid* is used instead. The standard *Cranberry juice drink with cranberry juice from concentrate* does not make a claim relating to vitamin C and lists vitamin C (rather than ascorbic acid) in the ingredients list. However, the nutrition declaration reveals that the standard product has 25% more vitamin C in it that the product purports to be the healthy option and which makes the nutrition claim.

What does the consumer understand by this? It is likely that the average consumer, having no special knowledge of alternate names for vitamins, would not

---

[17] Annex III of FIC.

[18] Annex III of FIC.

[19] A health claim is any claim that states, suggests or implies that a relationship exists between a food category, a food or one of its constituents and health as defined in Regulation (EC) No 1924/2006 of the European Parliament and of the Council of 20 December 2006 on nutrition and health claims made on foods.

[20] *High in vitamin C* claim requires a beverage to contain at least twice the amount required for a *Contains Vitamin C* claim by virtue of Regulation (EC) No 1924/2006 and Directive 90/496/EEC of 24 September 1990 on nutrition labelling for foodstuffs.

Fig. 13.2 JuicyCo's *No added sugar* c*ranberry juice drink sweeteners*

realize that the two terms are interchangeable in this instance. The consumer may think that the vitamin C content in the *Better Health* product comes from the cranberries in the produce and is not added as a separate ingredient.

13.3 Case Study 2: Supertreat Hot Dogs

Superfoods is a large European food company producing processed meat products, ready to eat meals and sauces to supermarket chains, wholesalers and food service via three different companies within its group. One of the companies, Supertreat, specializes in hot dogs and canned meat products and produces a wide range of

Fig. 13.3 Supertreat's *8 hot dogs made with mechanically separated chicken, in brine*

products. There are no European compositional requirements relating to the meat content of hot dogs. The name *hot dog* is regarded as a customary name which is defined as *a name which is accepted as the name of the food by consumers in the Member State in which that food is sold.*[21] Consumers will expect the product to have certain characteristics relating to flavour, texture shape and readiness to eat. Figure 13.3 shows Supertreat's cheapest product: *8 hot dogs made with mechanically separated chicken, in brine.* This product is sold through discount stores and corner shops and is often placed in supermarkets as a 'limited offer' product alongside the rest of the Supertreat range.

It is interesting to note the meat content of these hot dogs. *Meat* is defined at European level as *skeletal muscles of mammalian and bird species recognized as fit for human consumption with naturally included or adherent tissue.*[22] The definition continues to state that *products covered by the definition of 'mechanically separated meat' are excluded from this definition.* Mechanically separated meat (MSM) is *the product obtained by removing meat from flesh-bearing bones after boning or from poultry carcases, using mechanical means resulting in the loss or modification of the muscle fibre structure.*[23] MSM labelled as an ingredient is an important indicator of the quality of meat products, and its inclusion is often associated with cheap-quality products (European Commission 2010). Examination of the ingredients list whilst considering the legal definition of meat quickly leads to the conclusion that the product does not actually contain any meat (as defined) at all. This is shown in

---

[21] Article 2.2 of FIC.

[22] Annex VII Part B of FIC.

[23] Annex I of Regulation (EC) No 853/2004 of the European Parliament and of the Council of 29 April 2004 laying down specific hygiene rules for food of animal origin.

Table 13.1. Mechanically separated chicken is excluded from the definition of meat, and the other ingredients of animal origin (rind, fat and collagen) are not skeletal muscle and hence are not meat either.

The consumer may not expect the main ingredient of the product to be chicken, and they may not appreciate what 'mechanically separated' means. It is interesting how often casual conversations reveal that consumers expect inexpensive products of animal origin to have high-quality ingredients in them and are surprised that cheap products contain cheap ingredients.

The range of Supertreat products may cause further confusion for the consumer who is faced with a range of different types of hot dogs on the supermarket shelf, with variation in weight, unit price (i.e. price per 100 g) and meat and MSM from pork and chicken. Nine different Supertreat products are available in the supermarket, these being placed on the shelves in a block arrangement at around waist level, such that the consumer has a view of the complete range at once. Shelf-edge labels give the price of the product and also the unit price i.e. price per 100 g. Each product has full labelling in terms of the ingredients, nutrition label and other mandatory information.

Table 13.2 illustrates the information that may lead the consumer to select one product in the range over another. On inspecting the details of each product in the

Table 13.1 Calculating meat content of hot dogs

Ingredient of animal origin	Declared meat content	Actual meat content (as defined)
Mechanically separated chicken	62%	0%
Pork rind	–	0%
Pork fat	–	0%
Beef collagen	–	0%

Table 13.2 Net weight of packs and weight per item

	Description	Net weight (g)	Weight per hot dog (g)	Average weight of description (g)
	Hot dog			
1	8 hot dogs made with mechanically separated chicken in brine	184	23	
2	8 hot dogs in brine	184	23	
4	8 premium hot dogs in brine	184	23	27.48 (*Hot dog*)
7	6 New York-style hot dogs in brine	200	33.4	
9	10 New York-style hot dogs (vacuum sealed)	350	35	
	Jumbo hot dog			
3	6 jumbo hot dogs in brine	300	50	
5	8 premium jumbo hot dogs in brine	300	37.5	45.8 (*Jumbo hot dog*)
8	6 Jumbo New York-style hot dogs in brine	300	50	
	Giant hot dog			
6	8 giant German-style hot dogs in brine	720	90	90 (*Giant hot dog*)

Table 13.3 Details of the Supertreat range of hot dogs

	Front of label name	Drained weight (g)	Weight per hot dog (g)	MS chicken	Chicken	MS pork	Pork	Apparent meat content[a]	Actual meat content[b]	Price (euros)	Price per 100 g (cents)
1	8 hot dogs made with mechanically separated chicken in brine	184	23	62%	–	–	–	62%	0%	0.99	53.8
2	8 hot dogs in brine	184	23	58%	–	–	10%	68%	10%	1.10	59.7
3	6 jumbo hot dogs in brine	300	50	80%	–	–	–	80%	0%	2.75	91.7
4	8 premium hot dogs in brine	184	23	71%	–	–	8%	79%	8%	1.79	97.6
5	8 jumbo premium hot dogs in brine	300	37.5	45%	23%	–	7%	75%	30%	2.75	91.7
6	8 giant German-style hot dogs in brine	720	90	69%	–	–	14%	83%	14%	2.78	38.6
7	6 New York-style hot dogs in brine	200	33.4	81%	–	–	–	81%	0%	1.79	89.7
8	6 Jumbo New York-style hot dogs in brine	300	50	82%	–	–	–	82%	0%	2.75	91.7
9	10 New York-style hot dogs (vacuum sealed)	350	35	–	–	18%	67%	85%	67%	2.49	71.1

[a]What the consumer may assume on reading the label
[b]As per the EU definition of meat held in Annex VII of FIC

range, it can be seen that there is little correlation in terms of the names of products and weight, meat/MSM content and unit price.

Products are described in terms of their size as *hot dog*, *jumbo hot dog* and *giant hot dog*. The consumer would expect some form of progression in size, given the descriptors. Table 13.2 illustrates the net weight of packs and weight per item and is arranged according to the size descriptors.

Although there is a clear progression in the average of individual items described as *hot dog* (27.48 g), *jumbo hot dog* (45.8 g) and *giant hot dog* (90 g), there is variation in size for each of the given descriptors. *Hot dog* ranges from 23 to 35 g and *jumbo hot dog* from 37.5 to 50 g. Although this lack of standardization is not problematic legally, the inconsistency can cause confusion for the consumer. The net weight of pack varies for products described as *hot dog*, and for both *hot dog* and *jumbo hot dog*, there is variation in the number of items in the pack.

On inspecting the details of each product in the range, it can be seen that there is little uniformity in terms of the names of products and weight, meat/mechanically separated meat content and unit price.

Some of the products in the Supertreat range are described as *premium*. The consumer would understand that given a 'standard' product and a 'premium' product, the latter would be of a higher quality. In this case, it is likely that the consumer would expect a higher meat content or the meat used to be of a better quality. Table 13.4 compares the meat/MSM content of *hot dog* and *jumbo hot dog* with their *premium* counterparts.

There is little correlation between what the consumer would consider to be 'premium' ingredients of animal origin and those actually in the product. A comparison of *premium hot dog* and *hot dog* shows higher MS chicken content for the 'premium' product (71% as compared with 58%) but a lower pork content (8% as compared with 10%), with the actual meat content (as defined under EU law) being lower for the premium product. *Jumbo hot dog* has an even higher MS chicken content (80%) but no other meat nor MS meat content. *Premium jumbo hot dog* contains 30% meat as well as 45% MS chicken and does appear to be the most 'premium' of the four products. However, the consumer receives completely different products when choosing the standard or jumbo size option for each of the two varieties.

Table 13.4 Meat and mechanically separated meat content of standard and premium products

	Description	MS chicken (%)	Chicken (%)	Pork (%)	Apparent meat content (%)	Actual meat content (%)
2	8 hot dogs in brine	58		10	68	10
4	8 premium hot dogs in brine	71		8	79	8
3	6 jumbo hot dogs in brine	80			80	0
5	8 premium jumbo hot dogs in brine	45	23	7	75	30

Table 13.5 Meat and mechanically separated meat content of the Supertreat range

	Description	Apparent meat content (%)	Actual meat content (%)	Price per 100 g (c)[a]
	Hot dog			
1	8 hot dogs made with mechanically separated chicken in brine	62	6	53.8
2	8 hot dogs in brine	68	10	59.7
4	8 premium hot dogs in brine	79	8	97.6
7	6 New York-style hot dogs in brine	81	0	89.7
9	10 New York-style hot dogs (vacuum sealed)	85	67	71.1
	Jumbo hot dog			
3	6 jumbo hot dogs in brine	80	0	91.7
5	8 premium jumbo hot dogs in brine	75	30	91.7
8	6 Jumbo New York-style hot dogs in brine	82	0	91.7
	Giant hot dog			
6	8 giant German-style hot dogs in brine	83	14	38.6

[a]100c = 1€

Consumers may not be aware of what is meant by *mechanically separated meat* and may simply regard it as being part of the meat content of the product. Table 13.5 shows the meat and MSM content of products in the Supertreat range and compares this with the unit price of the product. The apparent meat content is the sum of the meat and MSM content and may be what the consumer understands to be the total percentage of meat as defined within the product. The actual meat content is the amount of legally defined meat in the product (i.e. not including MSM which is not within the EU legal definition of meat).

The actual meat content of *hot dog* varies from 0% to 67%. It is interesting to note that the unit price of the product with no meat in it is more expensive than the one with 67% meat content. The three varieties of *jumbo hot dog* all have the same unit price but vary from 0% to 30% meat content. *Giant hot dog* has the lowest price of all products in the range but the fourth highest meat content.

Table 13.6 shows the products in the range ordered from lowest to highest price. There is no correlation between apparent meat content and price, nor actual meat content and price.

Ordering the products by actual meat content as shown in Table 13.7 similarly shows no apparent correlation between price and actual meat content.

In deciding which products to buy, a value comparison cannot be made simply by looking at the unit price, since some of the lower-priced products in the Supertreat range have higher meat content and vice versa. The consumer is faced with a huge amount of information in making their choice and complex details that need to be considered in order to judge which product represents the best value.

Table 13.6 Meat and mechanically separated meat content of the Supertreat range ordered by unit price

	Description	Apparent meat content (%)	Actual meat content (%)	Price per 100 g (c)
6	8 giant German-style hot dogs in brine	83	14	38.6
1	8 hot dogs made with mechanically separated chicken in brine	62	6	53.8
2	8 hot dogs in brine	68	10	59.7
9	10 New York-style hot dogs (vacuum sealed)	85	67	71.1
7	6 New York-style hot dogs in brine	81	0	89.7
3	6 jumbo hot dogs in brine	80	0	91.7
5	8 premium jumbo hot dogs in brine	75	30	91.7
8	6 Jumbo New York-style hot dogs in brine	82	0	91.7
4	8 premium hot dogs in brine	79	8	97.6

Table 13.7 Meat and mechanically separated meat content of the Supertreat range ordered by actual meat content

	Description	Apparent meat content (%)	Actual meat content (%)	Price per 100 g (c)
7	6 New York-style hot dogs in brine	81	0	89.7
3	6 jumbo hot dogs in brine	80	0	91.7
8	6 Jumbo New York-style hot dogs in brine	82	0	91.7
1	8 hot dogs made with mechanically separated chicken in brine	62	6	53.8
4	8 premium hot dogs in brine	79	8	97.6
2	8 hot dogs in brine	68	10	59.7
6	8 giant German-style hot dogs in brine	83	14	38.6
5	8 premium jumbo hot dogs in brine	75	30	91.7
9	10 New York-style hot dogs (vacuum sealed)	85	67	71.1

13.4 Conclusions

In giving these two case studies, it is hoped that educators and students will be able to study comparable products available in their own country and open discussion on products that comply with all legal obligations but can still cause consumer confusion in respect of the information that is given. Many supermarkets and branded goods offer standard and 'healthier' variations of products, and the principles drawn out in the JuicyCo case study can be applied to any products in these ranges. Similarly, standard and 'budget' ranges can be compared in terms of ingredients and nutrition content. The Supertreat case study relied on the confusing number of products available across a range of closely related products of the same brand. Again, there are countless areas of food and nonfood products where the size of the range makes meaningful comparison for the consumer difficult.

References

Chandon P, Wansink B (2012) Does food marketing need to make us fat? A review and solutions. Nutr Rev 70:571–593

Elliott C (2014) Elliott Review into the integrity and assurance of food supply networks – final report. https://www.gov.uk/government/uploads/system/uploads/attachment_data/file/350726/elliot-review-final-report-july2014.pdf. Accessed 18th Apr 2015

European Commission (2010) Communication from the Commission to the European Parliament and the Council on the future necessity and use of mechanically separated meat in the European Union, including the information policy towards consumers. http://ec.europa.eu/dgs/health_consumer/docs/msm_report_20101202_en.pdf. Accessed 24th June 2015

Hui YH (2005) Handbook of food science, technology, and engineering, vol 3. CRC Press, London

Lähteenmäki L (2013) Claiming health in food products. Food Qual Prefer 27(2):196–201

Mephram TB (2000) The role of food ethics in food policy. Proc Nutr Soc 59:609–618

Oxford English Dictionary (2014) Online version. http://www.oed.com/. Accessed 16th Apr 2015

Rollin F, Kennedy J, Wills J (2011) Consumers and new food technologies. Trends Food Sci Technol 22:99–111

Temple NJ (2014) Food labels: a critical assessment. Nutrition 30(3):257–260

van der Merwe M, Venter K (2010) A consumer perspective on food labelling: ethical or not? Koers – Bull Christ Sch 75(2):405–428

Index

© Springer International Publishing AG 2018
R. Costa, P. Pittia (eds.), *Food Ethics Education*, Integrating Food Science and
Engineering Knowledge Into the Food Chain 13,
https://doi.org/10.1007/978-3-319-64738-8

Printed in the United States
By Bookmasters